RE-CRAFTING RATIONALIZATION

To Keith Weightman,
teacher and friend,
stopping off at the final clubhouse.

Re-crafting Rationalization

Enchanted Science and Mundane Mysteries

SIMON LOCKE
Kingston University, UK

Routledge
Taylor & Francis Group
LONDON AND NEW YORK

First published 2011 by Ashgate Publishing

Published 2016 by Routledge
2 Park Square, Milton Park, Abingdon, Oxon OX14 4RN
711 Third Avenue, New York, NY 10017, USA

Routledge is an imprint of the Taylor & Francis Group, an informa business

British Library Cataloguing in Publication Data
Locke, Simon.
Re-crafting rationalization : enchanted science and mundane mysteries.
1. Science--Social aspects. 2. Science--Public opinion.
3. Communication in science.
I. Title
306.4'5-dc22

Library of Congress Cataloging-in-Publication Data
Locke, Simon.
Re-crafting rationalization : enchanted science and mundane mysteries / by Simon Locke.
p. cm.
Includes bibliographical references and index.
ISBN 978-0-7546-7830-4 (hardback)
1. Science--Social aspects. I. Title.
Q175.5.L63 2010
306.4'5--dc22

2010035186

ISBN 9780754678304 (hbk)

Contents

Acknowledgements

Most of the material in this book has not appeared anywhere before, although some of the ideas and arguments have been aired in a variety of contexts before their specific formulation here. Particular acknowledgement needs to be given regarding Chapter 8, the central argument of which, although not the detailed development was published as 'Conspiracy culture, blame culture, and rationalisation', *Sociological Review*, 57(4), 2009, 567-85.

Thanks are due to Mike Hawkins for reading the draft and providing helpful comments and valuable discussion. Unpersuaded by the overall argument, I am nonetheless pleased to report that Mike has since become, as he put it, 'enchanted by the fact that [his] whole existence comprises an incomplete syllogism'. Welcome to the club, Mike! Now you can refuse to join.

Thanks and a great deal more are owed to Lorraine Allibone, who read and commented on parts of the draft, but more importantly provided unwavering support and encouragement throughout the period of writing, enduring labyrinthine expositions as I fumbled my way towards coherence – or as near as I can hope to get anyway!

Support of a more directly material form was provided by colleagues in the Sociology and Criminology Subject Group at Kingston University, who allowed me the relative luxury of a semester's sabbatical to complete the drafting of the manuscript. In these straitened times in the academy, it is heartening to think that there are still some who try to lift their gaze, if not quite to the stars, then at least a little above the bottom-line.

Thanks also to Neil Jordan at Ashgate for supporting the original proposal despite some unfavourable opinion and dealing with my queries with impressive efficiency and professionalism.

Introduction

This book is a theoretical study in the sociology of rationalization informed by empirical cases relevant to the public understanding of science (PUS). These meet in a concern with public (and popular) meanings of science (PMS) and the theoretical argument begins from the conviction that this is a fundamental concern of sociology that needs to be distinguished from the activites of the 'public understanding of science movement' (Gregory and Miller 1998) and its currently fashionable formula of 'public engagement with science and technology' (PEST) (Mellor, Davies and Bell 2008). For sociology, the proper matter of concern (Latour 2004) is to understand science in terms of its public meanings. This is in accord with the direction given by the founders of the discipline, especially Max Weber (1948b), whose thesis of intellectualist rationalization is precisely a statement about PMS (Locke 2001b). In the standard view of Weber's thesis, science and technology produce a monolithical condition of disenchantment in which the world is oriented to as knowable in principle though not necessarily known in fact. This then is about the *meaning* of science not its 'understanding' or 'engagement' in so far as these are taken to imply acceptance of the institutional authority of the established scientific community to dictate over 'matters of fact' (Shapin and Schaffer 1985, Wynne 1995, 2008, Irwin and Wynne 1996, Lock 2008).

However, the further theoretical argument of this book is that the standard view of rationalization articulates only *one version* of PMS that can be extracted from Weber's analytical framework. To be sure, it is one that Weber himself accentuated, but he did so as part of an argument concerned to emphasize what he saw as the appropriate vocation or ethic of conduct of the scientist (Lassman and Velody 1989). As such, he was necessarily arguing against alternative views and to do so adopted the terms of a formal rhetoric of public self-presentation used by scientists. There is an irony here, however, as Weber's own thesis allows for the alternatives he was arguing against as possible versions of science that arise from the developmental logic of intellectualist rationalization, or the unfolding discourse of the charismatic. These alternatives can be recovered through re-crafting rationalization to show that it allows for other rhetorical representations of science, ones that play crucial roles in scientists' internal argumentation and their external self-presentation. In respect of the latter, of especial importance are *enchanted* representations that are central to the popularization of science and hence its public meaning. However, the modes of internal argumentation are also significant as with its growing institutionalization they provide resources of public criticism of science.

This is not, however – and contrary to a growing chorus (Tiryakian 1992, Lee and Ackerman 2002, Lee 2003, 2007, 2008, 2009, Partridge 2004, Foltz 2006,

Possamai 2006) – an argument about *re*-enchantment in the west, because science has *always* been informed by the charismatic, notably in its 'grand metanarratives' (Lyotard 1984). It follows from this that PMS is characterized not by a singular condition imposed on people without their choice or involvement – we are not and *have never been* simply or *just* disenchanted – but by a central dilemma over how science is to be understood in this respect that involves active thinking and arguing (Billig 1996). Nor is this a once and for all matter, but an ongoing concern to be continuously re-thought and re-argued over. Disenchantment is then a *question* for us, but it has never been the only possible answer because it is countered by enchanted science.

In addition, this dilemma is cross-cut by a second, moral dilemma as to whether disenchanted or enchanted versions of science provide positive or negative prospective solutions to the problem of human suffering. This gives rise to the question of whether science produces a uniform, monolithical condition or one that is multiple and polyvalent – a closed or an open society (Popper 1966). Taken together, these dilemmas constitute articulations of the unfolding discourse of the charismatic that define a four-fold framework of ideal-typifed directions of development of intellectualist rationalization in modern western culture – or so I shall suggest.

To get to this point, I begin in Part I with the background theoretical work needed to re-craft rationalization. In Chaper 1, I make the case for PMS taking the reader on a bit of a ramble over the terrain of PUS with passing directional finders to PEST. This is not intended as a review of current research, assessment of the state of the field (Lock 2008) or proposal for scientific governance (Fuller 2000, Wynne 2008), but only as an argument for PMS grounded in a perspective I call *rhetorical sociology*. My route to this might seem somewhat oblique as it cuts through territory now often seen as the province of cultural studies, but to me remains part of the broad vista of the sociology of knowledge (McCarthy 1996, Chaney 2002) in the sense given by Berger and Luckmann (1967: 26) as 'everything that passes for "knowledge" in society'. Thus, science is viewed as an available cultural resource (Barnes 1974) in the social construction of meaning in everyday life, a term itself requiring some explanation to distinguish my take from others (Irwin and Michael 2003, Bell 2006, Michael 2006). Here, I draw on Miller and McHoul's (1998) 'ethnomethodologically inspired cultural studies' (EMICS), but critically reconfigure this through a rhetorical view of representation and ideology drawn from Potter's (1996) view of discourse, Billig's (1996) rhetorical psychology and Billig et al.'s (1988) thesis of ideological dilemmas. Thus, via a discussion of Miller and McHoul's use of membership categorization analysis (MCA) in respect of parapsychological phenomena and science fiction, I arrive at the conclusion that PMS is centred around a constitutive dilemma between the ordinary and the spectacular – or disenchantment and enchantment.

This then leads directly to Weber's rationalization thesis, the topic of Chapters 2 and 3. In Chapter 2, I begin to re-craft rationalization by recovering the range of alternative versions of science and its public meaning that are potentially implicated in Weber's analytical history of intellectualist rationalization as the unfolding discourse of the charismatic. This involves returning to the roots of the thesis in his 'philosophical anthropology' (Poggi 2006) concerning the human response to the existence of suffering through theories of salvation, or soteriologies, elaborated into ethics of conduct to inform specific ways of life (Shafir 1985). From a discussion of this thesis as developed in Weber's (1948a, 1948c, 1951, 1952, 1958, 1965, 1976) studies of world religions and culminating in his (1948b) ideal-typified model of the vocation of the modern scientist, I argue it is possible to extract alternative possibilities for both the scientist and PMS than solely the disenchanted instrumental reasoner Weber accentuated. Two things are stressed: that there is a disjuncture in his notion of 'formal rationality' between 'theoretical mastery' (abstraction of the charismatic) and 'calculation of means' (instrumental action); and that scientists' discourse is characterized by references to what he called 'substantive rationality' and 'irrationality', that is ethics of conduct drawn from other 'value spheres' such as the political and economic. The first opens the way towards enchanted views of science, while the second points toward resources of critical argumentation utilized in scientific controversies. Both are apparent in PMS, some support for which comes from a brief discussion of two currently fashionable accounts of the apparent growth in public critique of science, the 'condition of postmodernity' (Lyotard 1984) and 'risk society' (Beck 1992). I argue, however, that these two theses are *articulations* of the dilemmas of PMS rather than effective *analyses* of them.

This is brought out further in Chapter 3 where the alternative versions are elaborated. So, I continue to re-craft rationalization by linking Weber's distinctions between 'formal rationality', 'formal irrationality', 'substantive rationality' and 'substantive irrationality' with features of scientists' discourse, to argue that these refer to 'standardized verbal formulations' (Mulkay 1993: 71) that make up the available means of persuasion, or the rhetoric of science (Gross 1996). Thus, 'formal rationality' describes a formal, public rhetoric associated with disenchantment combining 'theoretical mastery' with 'calculation of means' (roughly rationalism and empiricism). 'Formal irrationality' refers to cognitive and methodological errors, features of scientists' discourse used in controversies to discount opposing views (Gilbert and Mulkay 1984; see also Collins 1985) and as resources of boundary-work (Gieryn 1995). Likewise, 'substantive rationality' and 'irrationality' refer to other resources of error accounting involving social interests and personal factors respectively (Locke 1999a). These techniques of argumentation have become taken up publicly as expert debate has become more prominent. In addition, however, enchanted images of science are used for purposes of popularization. Thus, the final step in re-crafting rationalization involves bringing out the disjuncture between 'theoretical mastery' and 'calculation of means'. I link this to Whitehead's (1974) identification of an ambiguity over disenchantment as involving both the elimination of magic through theoretical abstraction and universalization, and the restriction of acceptable modes of experience to an instrumental ethic of conduct. It is only the latter that leads to science displacing the charismatic – the version accentuated by Weber; the former, however, runs in the opposite direction towards science being informed by the charismatic – enchanted science. This is seen in the popularization of science as analyzed by the rhetorician Thomas Lessl (1985, 1989), who highlights scientists' use of synecdoche in forging cosmic connections between the human and the universal order involving what I call the construction of an *enchanted self*.

Having set out this theoretical basis, re-crafting continues through application to empirical cases, beginning in Part II with enchanted science in the contemporary public sphere. Here, I follow Whitehead's argument that the charismatic will continue its reach towards transcendent comprehension through a focus on symbolization seeking to overcome divisions between science and religion, but potentially running in contrasting directions towards a revived fundamentalism and a renewed occultism. These developments are apparent in the discourses of creationism/ intelligent design (ID) and Scientology respectively. Thus, in Chapter 4, I present an analysis of the use of synecdoche in the discourse of the British Creation Science Movement to show how the rhetoric of science is taken up and critically reworked to support a 'literal' reading of the Christian Bible. In a detailed analysis of one text concerning the evidence for ID in the workings of the human body, I look at how readers are invited to see themselves enchantedly in a discourse that syncretizes the biblical Word with the scientific world.

Then, in Chapter 5, I look at the use of the same kind of rhetoric in the writings of L. Ron Hubbard, the founder of Scientology, although turned to a quite contrasting end. I frame this analysis through discussion of the discourse of 'reflexive spirituality' (Besecke 2001) arguing that the view this gives of rationalization as involving a dynamic interplay – or *discursive syncretism* (Locke 1999a) – between religion and science was just that anticipated by Whitehead. Applying this to Scientology shows that the standard sociological view of this as a product of a modernity imbued with a disenchanted 'spirit' (Wilson 1990) is not only partial but overlooks the fact that Hubbard was actively critical of the version of science this view adopts, advocating an alternative that supports charismatic reaching beyond the merely instrumental. Thus, his writings, like those of creationists demonstrate that not only are *both* versions of science publicly available, but also that PMS involves active arguing over which meaning is preferred.

This comes out more fully in the discussion of science fiction/ fantasy in Chapter 6. Again, I follow Whitehead to argue that science fiction constitutes a cultural arena in which the charismatic reaching that informs science – the 'magic that works' in the words of John Campbell Jr – provides rocket fuel for the modern 'literary imagination'. I link this to Colin Campbell's (1987) thesis that the root of the 'Romantic ethic' derives from Puritanism providing a critical counterpart to the disenchanted instrumentalism of the work ethic. Thus, the dilemma between enchanted and disenchanted versions of science, and the moral dilemma over their utopian or dystopian implications define central narrative tensions informing science fiction as pithily expressed by Aldiss (1986: 26): 'hubris clobbered by nemesis'. To illustrate this, I refer to the case of superhero comics, which, whether considered science fiction or fantasy (Clute and Nicholls 1993), are a popular cultural form where the meaning of science is actively elaborated and interrogated (Reynolds 1992, Locke 2005). Moreover, in the phenomenon of *continuity*, superheroes also demonstrate the continuing process of rationalization within the aesthetic sphere as anticipated by Weber (1948b). But he did not anticipate the paradox of consequences that this would result in the fantastical becoming the basis of an alternative cosmology that seeks to transcend the division between 'fact' and 'fiction' by comprehending both as symbolic, magical constructs – a vision expounded in Alan Moore's 'science-heroine', *Promethea*. Thus, like 'reflexive spirituality', superheroes articulate the reaching of the charismatic for transcendence through symbolization.

Superhero continuity also illustrates something else: the ethno-method of the documentary method of interpretation (Garfinkel 1967) as a feature of mundane reasoning (Pollner 1987). This is both form and manifestation of rationalization – how rationalization is actively done – and in Part III, I move on to explore this further in what I call *mundane mysteries*. This refers to a paradox of consequences arising from the configuration of mundane reasoning under conditions of disenchantment. Disenchanted mundane reason pursues a demystified and definitive version of the world, but it results in the opposite – increased mystery. This arises partly because the means of error accounting employed by scientists enables the perpetuation of controversy potentially forever (Collins 1985, Billig 1996: 93-111) and partly because the disenchanted orientation of mind stimulates active enquiry into phenomena that itself generates mysteries. In addition, the former provides resources for the latter enabling alternatives to established scientific orthodoxy to be sustained, the outcome of which has been a growth and thickening of 'fringe' or 'marginal' science (Wallis 1979). In Chapter 7, I illustrate this by reference to the discourse of the followers of the early twentieth century collector of 'damned data', Charles Fort. Fortean discourse is substantially informed by the scientific fringe – including ufology, cryptozoology, parapsychology and alternative archaeology – although it takes in a wider sweep of alternative versions of the world. Thus, it both draws upon and is critical of established science (Parker 2001). Specifically, I show that Forteans draw on the resources of scientific error accounting by reference to *technologies of detection*, *psychologies of perception* and *sociologies of conception*. However, they also construct boundary demarcations to distinguish themselves from both 'believers' and 'debunkers' – as 'sceptics' but not 'skeptics with a "k"' – and through this strive to maintain discursive openness by sustaining the possibility of 'mystery'.

One form of sociology of conception that has a niche within Fortean discourse presents an especially encompassing reach – conspiracy discourse. I give this closer attention in Chapter 8 because it is particularly troublesome for professional sociology being a form of 'social and cultural theory' (Bell and Bennion-Nixon 2001: 149) with a similar structure of argument (Latour 2004: 229; see also Parker 2001). To make sense of this, I argue that conspiracy discourse is a form of *blaming of culture* in the terms of disenchanted mundane reasoning that provides a prospective account of suffering. Further, I suggest that the human sciences ideal-typically articulate the forms of moral accounting found in disenchanted mundane reasoning. Thus, the parallel between conspiracy discourse and sociology can be understood in that both are articulations of the available prospective soteriological potentials in an ostensibly godless world.

Finally, in Chapter 9, I pursue the analysis of moral accounting in disenchanted mundane reasoning into one further case of 'mystery' in which the scientist has been ascribed a particularly prominent role – Jack the Ripper. Here, I return to the discussion of MCA in Chapter 1, to propose this as a valuable technique for studying the category 'scientist' as both a source and focus of moral reasoning (see also Winiecki 2008). Through discussion of examples from Sacks (1995a), I argue that membership categorization devices (MCDs) employ incomplete syllogisms that are characteristic of what rhetoricians call enthymematic reasoning (Aristotle 1946, Billig 1996: 131-32, Gross 1996). Like enthymemes, MCDs are only probabilistic in the qualities or 'predicates' (Hester 1998) ascribed, or 'bound' to categories, which accounts for them being both potentially disputable (Silverman 1998) and providing invitations to speculative invention or 'making new knowledge' (Sacks 1995a: 42). I illustrate this through a reinterpretation of data taken from newspaper reports of the time (1888) about Jack the Ripper (Walkowitz 1992). Particularly prominent in these reports were speculations that the Ripper was a 'mad' or 'bad' doctor, understood as a man of science. I argue that these speculations show that the 'scientist' is both a resource of new moral categories and positioned within the existing moral order (Jayyusi 1984), such that the quality of 'instrumentalism' is both bound to the category of the scientist and bound to no good. Given that the depiction of Jack the Ripper as a 'mad doctor' has remained a recurring characterization in popular culture (Frayling 1986) and that mad scientists have continued to proliferate and hyperbolate (Skal 1998), this suggests that, contrary to the standard view of rationalization, instrumentalism has not displaced value-rationality, but rather the scientist qua instrumental actor is subject to mundane moral approbation.

This is further indication that the meaning of science is a matter of public and popular argument. Accordingly, modern culture has invented a diversity of images of the scientist and the activities bound to this category that run the gamut from heroic saviour to Faustian devil with numerous points between (Haynes 1994). This should not be over-simplified to a two-dimensional 'flip-flop' between extremes of good and bad (Collins and Pinch 1993) and still less reduced to a single 'pop' caricature (Basalla 1976) or one-dimensional instrumentalism (Marcuse 1964). PMS is vastly more resourceful than either of these wisdoms propose, containing a complex mix of disenchanted and enchanted representations of science, offering and subject to moral assessment as prospective solutions to the problem of human suffering in which the discourse of the charismatic continues to extrapolate and elaborate in its perpetual seeking for transcendent comprehension ... at least until the ironic cackle of the sociological anarchist breaks the spell!

PART I
Re-crafting Rationalization

Chapter 1
From PUS to PMS: Adventures of an Unhappy Acronym

PUS

> **pus** *n.* the yellow or greenish fluid product of inflammation, composed largely of dead leucocytes, exuded plasma, and liquefied tissue cells. (Collins 1994: 1260)

PUS pronounced as separate letters has another meaning: it is the widely used acronym within the British academic community for 'public understanding of science', the most recent wave of concern with which began in the UK during the mid-1980s (Royal Society 1985) continuing to the present albeit often under the rubric 'public engagement with science and technology' (PEST) (Mellor, Davies and Bell 2008). PEST supplanted PUS because of various problems associated with the so-called 'deficit model', assuming a top-down, unidirectional and monological relationship between knowledgable scientists and ignorant public (Wynne 1991, 1995, Irwin and Wynne 1996, House of Lords 2000), although the extent to which anything has really changed has been the subject of recent critical discussion (Wynne 2008; for a review see Lock 2008). In any case, PUS is not and has never been solely confined to the deficit model (Ziman 1991, Gregory and Miller 1998) and here the above definition of 'pus' is strangely appropriate. The thing is everybody knows what pus is, but it is unlikely that many of us, if asked, would trot out anything like this definition with its marked technical-scientific terminology even if we happen to know what leucocytes, plasma and so on are. In some contexts or circumstances, notably those of a medical or educational nature (or maybe for a certain type of quiz) such a definition and such an understanding might be appropriate or necessary, but for most ordinary, everyday circumstances it is not. Does this mean we do not understand pus? Of course not – the likelihood is we have all had some practical experience of it. But herein lies the conundrum of PUS: we, the general public, both understand science and do not understand science; in an important sense – or more correctly in a range of important senses – we know what science is, but in other perhaps equally important ways we know nothing about it whatsoever (Michael 1992).

The problem with the deficit model is that it tended to accentuate the latter and do so from the perspective of aggrieved (mainly natural) scientists worrying over the perceived decline in public support for science. Scientists, being the knowledge workers they are saw this as being in good part a problem of ignorance. However, for other (mainly social) scientists the issue put this way was wrongly conceived. The question was not so much one of what people do *not* understand, but what they *do* understand and what, if anything, science has to do with it – where does science fit in people's commonsense outlook on the world, in 'everything that passes for "knowledge" in society' (Berger and Luckmann 1967: 26)? 'Understanding', in other words, was understood in a very different way (Irwin and Wynne 1996; see also Wynne 1991, Ziman 1991, Lévy-LeBlond 1992, Gross 1994b, Yearley 1994). So, if it is in fact the case that members of the public are not able to define 'pus' in anything like its technical sense, then this tells us in a way nothing of particular interest – although that some social groups might think they ought to be able to do so is very interesting and tells us vastly more about them than about the public.

Something of what it tells us is again apparent from the above definition. Although other definitions of 'pus' are available, this one is notable for its technical-scientific vocabulary. This is an indication of the extent to which the scientific community has become legitimized as a key institutional source of public meaning-making. The specialized discourse of this one particular social group is presented, at least in the formal context of a publicly defined means of meaning (a dictionary), as providing the general meaning of the word for all social groups. Thus, to an extent, the scientific community is invested with the right to control meaning; it has a publicly recognized sanction or 'category entitlement' (Whalen and Zimmerman 1990) to do so. Most especially it presumes the right to define the terms of its own discourse (Sharrock 1974). Given this it is perhaps unsurprising that scientists might set out with the assumption that their meaning is the right meaning and if members of the public are unaware of it or use a different meaning then this is a measure of ignorance and justified cause for concern.

However, having such formally recognized public authority does not necessarily mean that either the right to make meaning or the meanings themselves are always accepted or go unchallenged. As the example of 'pus' suggests many terms employed in scientific discourse are derived from ordinary, everyday words on which to an extent the specialized language of science is parasitical. Scientists may attempt to re-define these words in their own vocabulary, but there is no guarantee the new definitions will be accepted or that existing meanings and uses will disappear (Messeri 2010). Even if a new meaning is adopted, it will always be put to use in specific ways in specific contexts for specific purposes and its meaning will begin to transmogrify accordingly – rather like how PUS became PEST. The general point here is the rule about rules: no rule can cover the rules of its own application (Collins 1985; see also Winch 1990, Barnes, Bloor and Henry 1996: 46-80). In respect of words, their meaning comes from how they are used within a culture or 'way of life' (Williams 1976). Thus, however much science may have formal public sanction to define terms and however much scientists may attempt to control their meaning, meanings will always slip away because how the terms will come to be used cannot be controlled. Or, to turn this around, the attempt to police meaning is the attempt to police ways of life; so when scientists talk about public misunderstanding they are also talking about public unruliness (Lock 2008).

At one level this is about an attempt to bolster the institutional authority of science (Wynne 1992, 2003, 2008) within which 'public' refers primarily to formally constituted bodies, systems or apparatus that are often taken to be dominant over if not coterminous with society (Erickson 2005). Such an institutional public has been a focus of concern regarding the governance of science (Fuller 2000) in what are taken to be changing social conditions of knowledge production (Nowotny, Scott and Gibbons 2001) and how it relates to other institutional contexts such as the economy, social policy and law (Irwin and Michael 2003, Yearley 2005), concerns given added impetus by the recent spate of issues involving 'risk' and the environment. However, 'public' can take other meanings; like 'science' and 'understanding' it is an *envelope word* into which a range of different, even contrary meanings can be shoved (Cooter and Pumfrey 1994, Michael 2009). These words then have about them a certain kind of vagueness (Potter 1996: 166-69) lending themselves to enthymematic rhetorical reasoning (Locke 2002 and see Chapter 9). Consequently, they can be used to support a range of different interests by being filled out or spun to suit. But it also follows that there is no guarantee they will be understood or accepted in the intended manner since the incomplete syllogism of an enthymeme can be completed in different ways by different audiences. People can read many different things into these words, with the further consequence that they might assume they are talking about the same thing when they are not – and that can lead to trouble (cue the 'Science Wars': Gross and Levitt 1994, Ross 1996, Gross, Levitt and Lewis 1997; for an attempt at 'science peace' see Labinger and Collins 2001).

Another way of putting this is to say these words are polysemic, a term used in cultural studies to refer to the multivocality of texts (Hall 1980, Hüppauf and Weingart 2008). There is a peculiarity about this in the case of scientific discourse, because this is marked by a particular kind of definitive form – a 'rhetoric of no-rhetoric' (Beer and Martins 1990: 171). The technical discourse of science purports to present us with the world simply as it is unmediated by human contact, a magical denial of its own discursive character (Locke 1999a: 107). Thus, the formal, institutionalized public discourse of science tends to appear monovocal and lacking the interpretative multiplicity of polysemic texts. There is an attempt to close down meaning and establish a single, hegemonic meaning creating what Collins (1985) calls a 'ship-in-the-bottle'. But the strong view in cultural studies – like the equally strong view in the sociology of scientific knowledge (SSK) (Mulkay 1985, Woolgar 1988, Ashmore 1989) and the 'constitutive' view of rhetoric (White 1987; see also Gross 1996) – is that all texts are polysemic because they are produced by and received in discursive environments with competing voices at work within them (Fiske 1998; see also Bakhtin 1981). Maintaining hegemony involves continuous struggle as the tricks used to place the ship in the bottle can become exposed. This is readily apparent in the case of scientific texts because they always have an argumentative character however compacted and indirect this might appear (for example, Myers 1990, 1997). In Vološinovian (1973: 21) terms, they have an 'evaluative accentuation' or what might be called *rhetorical valency*.

For example, it is often said in popular books these days that science is about testing theories (for example, Eldredge 2000). This is a fairly recent development; in the nineteenth century the standard public form to follow was induction (Gross 1996, Campbell 1997) which shows that scientists present different public images of science at different times and places (Gieryn 1983, 1999). Further, what 'testing theories' might mean for the practice of scientists is open to interpretation,[1] but in so far as it describes a feature of the genre of scientific reports (Bazerman 1988) then they will take an argumentative form either in support of or against one or other particular theory. In so doing they will necessarily imply an opposing viewpoint leaving open the potential for that alternative to be further developed (Billig et al. 1988). If this were not so it is difficult to see how scientific debate could occur, but it does even over something as supposedly fundamental as evolution – and without needing to raise the spectre of creationism (or intelligent design) as seen in the debate between Neo-Darwinian gradualism and 'punctuated equilibria' (Eldredge 2000). This point is implicitly recognized by those who argue that creationism should not be discussed so as to starve it of publicity; on the other hand, the counterview argues equally strongly that if it is ignored it cannot be refuted (see also the case of 'forbidden archaeology' – Cremo 1998). This points to one of the dilemmas of popularization, more on which shortly.

But even on a more prosaic level the technical discourse of science is polysemic. Consider again the definition of 'pus'. Although not strictly 'scientific' since it comes from a general dictionary, it does have something of this character in referring to certain technical objects of scientific discovery and, as noted, it does so with the definitiveness of scientific discourse. Even so, we do not have to look too hard to begin to spot some cracks in the neat finish. For example, the 'fluid' is said to be 'yellow or greenish', so what colour is that exactly? It is also said to be 'composed *largely* of dead leucocytes (etc)', so what else is in there? Do we know? The fluid is said to be the 'product of inflammation', which seems to suggest pus is caused by inflammation, but should it be the other way around? We can go further and think about the imagery in the text, its metaphors redolent of ancient elemental discourse of water ('fluid'), fire ('inflammation') and earth ('tissue'), and note the irony that the word 'cell' is derived from the monastic life of the Middle Ages, a sign of science's religious roots (Merton 1970, Fuller 2007). Or if we are informed by another type of science, we can draw on contemporary social theory to consider how the text embodies heterogeneous 'actants' (Latour 2005) not only in its melding of these multiplicitous discursive fragments, but in their

---

1 This is not the place to revisit the relevant debates in the philosophy of science, so suffice it to say the pertinent question is, if scientists do test theories, are they their own (Popper 2002) or those of others (Feyerabend 1993)?

instantiation in the material object of the dictionary to be potentially positioned in a myriad of diverse networks of 'co(a)gents' (Michael 2000, 2006). And so on.

But the fact that I can pick a bit at a dictionary definition might not seem to tell us much about PUS – after all, my academic membership category (see below) attests to me supposedly being some kind of 'expert' (if not quite a virtuoso – Fuller 1997) at rhetorical nit-picking. Here, however, we come back to the multiple meanings of 'public'. In addition to referring to a formal institutional order of discourse, 'public' has two other prominent meanings in the debate about PUS (see also Einsiedel 2000). One refers to the mass media as the principal outlet by which, supposedly, society at large holds a conversation with itself as expressed in the idea(l) of the 'public sphere' (Habermas 1989, Cooter and Pumfrey 1994, Gregory and Miller 1998). The other refers to everybody within society, all of the ordinary people who constitute its membership, often downgraded in the discourse of PUS to 'non-scientists' (overlooking that scientists are themselves part of the public to disciplines outside their own specialism) though sometimes politically upgraded to 'citizens' (Irwin 1995, Durant 2008: 18). These two meanings significantly overlap in another envelope word, 'popular' (Reinel 1999), which calls for a slight digression.

P.T. Barnum Meets Dr Frederic Wertham

As is more fully developed in later chapters, a central argument of this book is that the multivocality of scientists' discourse has profound implications for PUS. The issue needs some immediate discussion however, because the opposite view has been taken chiefly on the grounds that the technical discourse of science effectively accomplishes closure of controversy (Collins 1985, 1988). Since to a considerable extent only this discourse is accessible to 'non-scientists' as they (we) are routinely excluded from 'science-in-the-making' (Latour 1987), then it is this finalized version of reality that we accept – at least most of us most of the time. We only see the spruced up ship-in-the-bottle, so we cannot discover the tricks used to put it there even if we think there might be some. I call this the *P.T. Barnum postulate* after the remark attributed to the great nineteenth century American showman, 'there's a sucker born every minute'. Barnum peddled purported 'freaks' and 'marvels' that the supposedly ignorant and gullible public accepted at face value and I discuss some such marvels, or *mundane mysteries*, in Chapter 7 drawing quite contrasting conclusions.

In a nice irony this postulate unites the natural scientific proponents of the 'deficit-cum-engagement' model with some of their sociological opponents in the erstwhile Science Wars, as can be brought out by considering the parallels between PUS discourse and the long-running debate about mass culture. This debate also united some odd bedfellows, notably cultural conservatives like Malcolm Arnold and Q.D. and F.R. Leavis, with critical theorists like T.W. Adorno and Max Horkheimer (Storey 2003). Adorno also happened to be an acquaintance of Frederic Wertham, a German emigrant psychiatrist based in New York who is most remembered – if not always favourably – for his role in the 'anti-comics campaign' of the early 1950s (Wertham 1955). This had a major impact on both sides of the Atlantic leading in Britain to legislation (the 'Children and Young Persons (Harmful Publications) Bill' of 1955) and in the United States to comics publishers voluntarily submitting to censorship in the form of the Comics Code Authority (Barker 1984, Nyberg 1998, Lent 1999, Hajdu 2008). A central role in this debate was played by ideas about science and the public.

The view of the public at work in the anti-comics campaign can be gleaned from the British Bill, which states in part the following:

> This Act applies to any book, magazine or other like work which consists wholly or mainly of stories told in pictures (with or without the addition of written matter), being stories portraying–
>
> a. the commission of crimes; or
>
> b. acts of violence or cruelty; or
>
> c. incidents of a repulsive or horrible nature;
>
> in such a way that the work as a whole would tend to corrupt a child or young person into whose hands it might fall (whether by inciting or encouraging him to commit crimes or acts of violence or cruelty or in any other way whatsoever).

The worry expressed here is about adverse media 'effects' on parts of the general populace judged to be 'vulnerable' (Barker 1984) and it is this that unites cultural conservatives and critics alike, albeit for quite opposing reasons: conservatives fear social collapse and revolution; critics, social stability and passive acquiescence (on the parallels between media effects and ideological analysis, see Barker and Brooks 1998: 83-108). Both are forms of what I call *victim sociology*, some further examples of which are encountered in later chapters. The parallels with the Science Warriors are clear: those who worry that the public are ignorant about science speak in the same discursive terms as those who worry that an ignorant public are vulnerable to adverse influences from the media; meanwhile, those who worry that the public are ignorant of science-in-the-making speak in the same discursive terms as those who worry that a passive public are held in thrall by the pre-digested ideological formulae of the culture industry. *Both* speak, however different their inflection, with the voice of P.T. Barnum.

This is further shown by the role played by science in the Wertham case. He has often been criticized for poor social scientific procedure as he did not construct a systematic sample or use a control group in studying comics and their readers (Barker 1984, Brown 1997). Evidently then there are grounds to warrant the worry that ignorance about science can lead to poor policy-making. The anti-comics campaigners – including many ordinary people, such as parents and retailers, as well as politicians – were evidently ignorant about social science and hence vulnerable to being misled by a purported expert who it seems based his conclusions on inadequate evidence. On the other hand, however, as comics fans have since taken up this kind of criticism in a bid to defend the comics, perhaps we should take heart that at least some members of the public are able to get it right – not so much the P.T. Barnum postulate perhaps as the Honest Abe dictum, 'you can't fool all of the people all of the time'.

Unfortunately, however, the matter does not rest there, because Wertham's procedures have been defended on the grounds that they need to be understood in relation to his analytical perspective, which was based on a non-positivistic conception of social science influenced by Adorno's critical theory and not out of keeping with forms of ethnography used in contemporary cultural studies (Nyberg 1998: 86-96). Thus, whether we view his research as properly scientific or not depends on whether or not we accept this perspective. If we do, then our first inferences about public understanding would now be reversed: it is the campaigners who properly understood matters and the comics fans who show their vulnerable ignorance.

But there again the matter does not rest. Surely all this only goes to illustrate the second kind of concern above that the public are ignorant of science-in-the-making and if campaigners and fans alike had had this level of understanding from the outset then none of the debate would have occurred. They would have been aware of the unresolved and problematic status of Wertham's research or at least would have known better than to put much faith in the claims of any such expert. In effect, both groups are vulnerable and both have been misled all because scientific experts are given too prominent a role in public decision-making, which only leads to uncritical faith on the one hand and outright rejection on the other – a disastrous 'flip-flopping' of public opinion (Collins and Pinch 1993). Accordingly Wertham is on the one hand sanctified and on the other demonized; like so many images of the 'pop' scientist (Basalla 1976), he is either simple hero or villain, neither characterization justified because he was really only all too human.

Despite their evident differences however, the two views are united around a common concern with accuracy. From the first position what is considered important is that the correct finished version of science should be communicated to the non-scientific public (so, was Wertham right or wasn't he?). From the second position, however, what is considered important is that the correct view of science should be communicated, specifically that it does not really have finished versions (so, we don't know for sure if Wertham was right or not) (Durant 1993). This raises the topic of the 'public sphere' understood as a site of collective rational discussion within which the mass media are assumed to be crucially positioned. The example of comics is again helpful because as a mass medium it raises two entangled issues regarding science popularization and the wider debate about 'dumbing down'.

Bad Science, Conspiracy and Superman

From the first point of view, science has major contributions to make to rational public discussion through the provision of correct information and improved understanding of science's techniques of forensic analysis to weigh the quality of evidence and argument. However, there are also significant dangers of misinformation, distortion and over-simplification, perennial sources of concern regarding popularization as scientists' 'messages' are channelled through the noisome interference of journalistic agendas, to say nothing of journalists' 'ignorance' (Hilgartner 1990, Lewenstein 1995). As the media, epitomized by the unruly internet, have become increasingly prominent within the public sphere so this dilemma has become more pressing, the more so if popular culture is associated with the lowest common denominator – the already dumb with the even dumber. Or, as a seemingly endless line of books over recent years would have it, a sphere of 'weirdness' (Shermer 1998), 'mumbo-jumbo' (Wheen 2004) and 'voodoo' (Park 2000, Aaronovitch 2009), to say nothing of 'bad science' (Goldacre 2008). Hence, PEST as a hopeful compensation.

From the second point of view, however, PEST does nothing to address the real issues and is itself part of the problem. Here, the functions of popularization are seen as less about raising the bar of public rationality than raising barriers to public scrutiny, something to which scientists contribute by presenting science as definitively knowledgable and capable of far more than it can achieve. Thus, whether they intend it or not, the 'voodoo' comes from them as they appear to the public as whizzed-up shamen, high-priests of a modern technological wizardry with a gadget to solve every problem. But the magic does not come cheap; it takes time and money. So popularization is really about maintaining social position, ensuring jobs for the boys (and a minority of girls) and fending of competitors in the knowledge production stakes such as religion. The point of 'engaging' the public is to stop them being frightened off. For this simplification is necessary: it makes no more sense for scientists to present their back-stage rigging than it does for a stage-magician to expose their magic secrets. Science in public is as much a 'front-stage' performance as any other social act (Goffman 1990). It is not then journalists who are responsible for dumbing down, but scientists themselves concerned to win the public to their cause (Nelkin and Lindee 1995, van Dijck 1998) or gloss over cracks in the sheen of definitive knowing (Collins 1988, Gieryn and Figert 1990). The danger is not over-simplification but over-exposure, just the kind of exposure that the proponents of this view advocate in the conviction that the real contribution of science to rational public discussion is the capacity for self-critique that SSK in particular has advanced – providing this is kept within certain limits anyway (Collins and Yearley 1992). It is in this that the public needs to be educated; it is not their fault they are dumb as this is the outcome of an effective conspiracy of exclusion resulting from machinations at the core of the scientific community working to close down meaningful public debate.

That both these views are problematic, however, is shown by some consideration of superhero comics (see also Chapter 6). Elsewhere (Locke 2005), I described the traditional perception of superhero comics in the English-speaking world as 'thrice damned': as 'low' culture (even by the standards of popular culture); as dangerous medium (after Wertham above); and as an absurdly silly genre ('biff', 'pow' etc). Surely, then, here we have the dumbest of the dumb! And yet not only do superhero comics problematize the view that the mass media and popular culture are purveyors of 'bad science', they also provide grounds for doubting that the public are altogether ignorant of science-in-the-making.

As far as 'bad science' goes consider the superhero-in-chief, Superman. On the face of it what could be more absurdly scientifically inaccurate than a being capable of out-powering a locomotive, leaping tall buildings and racing faster than a speeding bullet? The physics and biology of such abilities are completely impossible (Gresh and Weinberg 2002) and this does not begin to address additional powers such as unassisted faster-than-light flying and travelling through distant space and time. Of course, Superman is only a fictional, indeed fantasy (Clute and Nicholls 1993) character and nobody is pretending his feats are really possible. Nonetheless, efforts have been made since his inception to invest him with some believability through proferring possible scientific explanations for his abilities (Siegel and Shuster 1997[1938]). Books addressing the 'science of superheroes' may have proliferated in recent years (Yaco and Haber 2000, Wolverton 2002), but there is nothing new for comics readers in this speculative sport. The details of the explanations have been modified over Superman's seventy year history, but one basic feature remains the same: he is an alien being. Might it then be possible, by some stretch of scientific plausibility that somewhere in the universe a species may have evolved which, although it may look like *Homo sapiens*, nonetheless is endowed with physical attributes akin to Superman's super-powers?

Before we can get anywhere with such an argument we first have to establish that intelligent life is possible on other planets. Famously a calculation to this effect has been made: the Drake Equation, worked out by scientists Frank Drake and Carl Sagan in the 1950s. According to Gresh and Weinberg (2002: 10) they came up with a high-end estimate of ten thousand planets with 'technologically advanced civilization' in our galaxy alone. This is some way from postulating the existence of an alien race of super-humans, but it might give us grounds for wondering if such a thing is not entirely outside the realms of possibility, in which case maybe Superman is not altogether such bad science. Then again, Gresh and Weinberg (2002: 9) cite other scientists who argue that Drake and Sagan heavily over-estimated the number of habitable planets. They, like many other 'space scientists' adopted the 'Principle of Mediocrity', which assumes '[e]arth isn't special, so there should be lots of other planets with life on them', but this has been called into doubt by 'what we've learned about astronomy in the past few decades' (Gresh and Weinberg 2002: 13). So, then, we seem to be back where we started: if we cannot even get off the ground with the possibility of intelligent alien life, then flying faster-than-light has no chance; it looks as though Superman is bad science after all. On the other hand, if we wait a few more decades maybe some more space scientists will come along and tell us differently again (see Fort 1996)!

The serious point is that what is and is not 'bad science' here is far from clear cut or straightforward. All the participants evidently are scientists; who then should the non-scientist believe? When even the experts disagree, how are the rest of us to decide? And given that expert disagreement is not so unusual and arguably increasingly publicly common, the question of whether the media present us with 'accurate' or 'inaccurate' science is also far from clear cut. This begins to look rather like the age-old strategy of blaming the messenger, a strategy designed to deflect responsibility and cover up the real source of the problem, which brings us to the second view.

The idea that professions are 'a conspiracy against the laity' goes back at least to Bernard Shaw (1932 [1911]), but even if true it must be a very bad conspiracy because the laity appear to be well aware of it. Superhero comics again provide evidence as it has become an established narrative feature, to the point of formulaic redundancy, to present super-powered beings as products of secret experiments undertaken by institutional authorities such as the government, military or big business. Needless to say, these experiments involve the use of advanced technoscience and as such continue the basic feature established with Superman that superheroes – and super-villains – are very often given a gloss of scientific and/or technological plausibility. Indeed, they are often scientists themselves whose experiments lead to their powers or backfire to produce unexpected monstrous outcomes resulting in characters who turn against them. A clear narrative parallel can be drawn with the Jewish legend of the golem, as well as the more temporally proximate and characteristically modernist story of Frankenstein's monster (Toumey 1992, Turney 1998a; see Chapter 6) and it is notable that many involved in the early comic book industry were both Jews and science fiction fans (themes found in Chabon 2000; see also Jones 2006, Kaplan 2008).

It might be objected that mad scientists producing monsters is hardly equivalent to the sophisticated exposure of subtle processes of social construction at work in laboratories and other arenas of science-in-the-making (Knorr-Cetina 1981, Collins 1985, 1987, Lynch 1985, Latour and Woolgar 1986, Latour 1987). But that depends on what we mean by 'social construction'. One meaning of this is 'social interests', as in the work of the Edinburgh School of SSK, which following Bloor's (1976) Strong Programme took these to be the principal 'social causes' of science (Barnes, Bloor and Henry 1996). Here it seems to me that the differences are more to do with the maintenance of academic boundaries from the (ignorant) public than the perspectives involved. As Latour (2004: 229) has remarked, the structure of argument between conspiracy thinking and 'critical' social theory is similar to the point of identity (see also Locke 1991), a parallel explored further in Chapter 8. However, 'social construction' may mean something more relativistic like Collins's (1983) erstwhile 'Empirical Programme of Relativism'.[2] Surely

2 Collins and Evans (2002) have since navigated to the 'third wave'.

we do not find in comics anything remotely like the notion of 'the interpretative flexibility of experimental data' with the attendant idea of the accepted version of reality being a social, rhetorical accomplishment? Perhaps not, but superhero comics are stuffed to bursting with multiple realities, different versions of history, alternative dimensions and possible futures, as well as playful interweavings of 'fact' and 'fiction' – the Fantastic Four, for example, have had a deal with Marvel Comics to produce comic book versions of their adventures since 1963 (Lee and Kirby 1963a). There is nothing new for readers of superhero comics in the idea of alternative science(s) and technology(-ies), however sketchy they might be in their details. This may not be 'social constructionism' in any academically virtuous sense, but it shows considerable flexibility in the view of science and its relation to reality and an implicit awareness that the boundary-line between 'fact' and 'fiction' is a moveable product of human imagining.

Am I then saying that comic book writers and readers are relativists and that we can expect them to support a critical science and technology studies' (STS) view of science, calling for more democratic institutional systems and the introduction of social constructionism in science education? No – for the simple reason that no conclusions about specific attitudes, perceptions and political and social commitments necessarily follow. But doubts should be raised about the adequacy of the meaning of 'the public' in the various deficit versions of PUS/PEST leading to questions about what kind of 'understanding' (or 'engagement') we are interested in of what kind of 'science'. To explain a little further what sense of these this book deals with, I now outline an alternative that I call, with yet another unhappy acronym, public (and popular) meanings of science (PMS).

Remi(cs)x: PMS

> **PMS** *abbrev. for* premenstrual syndrome ... *n.* any of various symptoms, including esp. nervous tension, that may be experienced as a result of hormonal changes in the days before a menstrual period starts. (Collins 1994: 1199, 1227)

Like the definition of 'pus', this definition of 'PMS' is also strangely appropriate and not just because it bespeaks a medicalization of the experience of women, transmogrifying an aspect of the ordinary lives of some people into a scientifically defined condition. More pertinent is its vagueness ('any of various symptoms') and its potential implications of irrationality ('nervous tension'), as these are both qualities often associated with the phenomena I intend to include under PMS.

The kinds of things I am chiefly interested in are often associated with so-called 'anti-science' (Holton 1992, 1993), which like the dictionary sense of PMS has the qualities of a 'syndrome', that is 'any combination of signs and symptoms that are indicative of a particular disease or disorder' (Collins 1994: 1564). 'Anti-science' encompasses a combination of signs and symptoms taken to indicate a supposed disorder affecting modern societies in which the advances of science are rejected in favour of various 'irrational' beliefs, including forms of religion both new and traditional, and a range of 'alternative' ideas of a 'superstitious' or 'supernatural' – broadly 'paranormal' – form. The way such beliefs are represented is not without parallel to the way women have been said to be positioned by modern science as an irrational 'other' (Haraway 1991, Harding 1991, Hess 1997). As a feature of its professionalization, formally rationalized science constructs a lay public as a similarly 'irrational' other as a means of contrastive definition, as seen in attempts to differentiate science from 'commonsense' (for example, Wolpert 1992; see also Pollner 1987: 6-12, Mellor 2003, Evans 2009). Hence, 'anti-science' is not a phenomenon so much as a resource of the 'practical sociological reasoning' (Garfinkel 1967, Frances and Hester 2004) of members of the scientific community used to accomplish social closure through boundary-work (Gieryn 1983) and to account for perceived loss of support.

This articulates the same dilemma encountered above regarding the media: on the one hand, the public is needed to maintain legitimacy as shown by 'public witnessing' being a foundational component of the 'literary technology' of science (Shapin and Schaffer 1985; see also Shapin 1990); on the other hand, exactly who this 'public' is needs careful management to maintain exclusivity (Lessl 1989, Shapin 1994, Gregory and Miller 1998, Reinel 1999, Broks 2006). 'Anti-science' implies the wrong kind of public. There is an irony in this, however, as it might be argued that far from being 'anti-science', the public(s) in question have learned the lessons of formal, institutionalized science rather too well for comfort. As such, they are an unintended outcome of scientific institutionalization, or in the Weberian terms to be explicated in Chapters 2 and 3, a paradox of consequences of the process of rationalization. However, to see them this way involves a re-crafting of rationalization to rescue Weber's thesis from the standard view adopted in much social theory. This stresses 'disenchantment' as the prevailing mental outlook of the ordinary person produced by the process of 'intellectualist rationalization' that Weber (1948b) associated with modern science. Effectively, this is an argument about PMS (Locke 2001b), but it assumes that the formal, public discourse of science is dominant and so neglects the 'informal' responses I take to be central features. The 'meanings' that PMS is concerned with, then, are circumscribed by this concern (compare Wynne 2003, 2008)[3] and my discussion is focused around the extent to which the meaning of science ordinarily available to people is limited to that presented by the standard view of rationalization.

A critical issue then is the meaning of 'ordinary'. Here, the notion of 'informality' as used in such literatures as the sociology of organizations and economic life (Pahl 1984, Harding and Jenkins 1989, Locke 2001c) provides a link to the wider

3 The point to stress being that, contrary to some (Collins and Evans 2002: 281), my position should not be conflated with Wynne's, although I share his concern that public meaning(s) of science should not be restricted to matters of 'expertize'. Nor should they be viewed as simply a matter of *quantity*, tucked away somewhere in metaphorical 'pockets' (inside a magician's cloak, perhaps), but of *conditions of possibility* (Locke 1999b).

meaning of 'public' mentioned above. Informality in these contexts refers to ways of doing things that are 'unofficial' or 'off the books', but nonetheless part of the normatively accepted order of everyday, ordinary social activity. They may not be accepted as right (in the 'formal' public sense), but they are widely accepted as things that pretty much everybody does to some extent or other, at some time or other. Some of these activities may be formally illegal, but most are not, although they may be 'deviant' when viewed from the formal version(s) of reality; more correctly, they raise questions about the application, the meaning-in-use of the formal rules, so that whether they are 'deviant' or not becomes a matter of how the rules are understood in, through, and by their application (Zimmerman 1971), and who is doing the defining. They may then be the focus of dispute: is this particular action correctly following the rule or not (see also Billig 1996)? In the terms of ethnomethodology, how is it made accountable (Garfinkel 1967)? In relation to science, then, PMS refers to the complex of 'informal', 'unofficial', 'deviant' versions and uses of science, and the accounts that are given of them that circulate in the social world and which might and often do work to make them appear 'formal', 'official' and 'not-at-all-deviant' (on the problems of the 'formal'/ 'informal' conceptual distinction, see Bittner 1974, Boden 1994). It is these accounting procedures and how we are to make sociological sense of them that are examined in the case studies of 'rejected knowledge' (Wallis 1979) in Parts II and III.

However, to claim that these accounting procedures circulate in everyday social life begs the question of what 'everyday life' means. Here, I draw on Miller and McHoul's (1998) 'ethnomethodologically inspired cultural studies' (EMICS), although I diverge from them significantly in a rhetorical direction (for other discussions of 'everday life' in relation to public science, see Bell 2006, Michael 2006). It is then tempting to call my approach REMICS, standing for 'rhetorically and ethnomethodologically inspired cultural studies', but this is a bit of a mouthful and *rhetorical sociology* is rather more apt. Rhetorical sociology takes from ethnomethodology a focus on the practical sociological reasoning visibly displayed in accounts about science, but views this in rhetorical terms as involving 'the available means of persuasion' (Aristotle 1946: 1355b, 26; see also Gross 1996) in two major senses: how and what versions of science are employed as suasive devices; and how these versions are used in the construction of alternative, 'unofficial' discourses. This practical reasoning is theorized through a re-crafted version of rationalization set out in Chapters 2 and 3. It is important to stress that 'rhetoric' is not used in any kind of pejorative sense and contrasted neither to 'reality' nor 'truth'. It is intended in the sense employed in the stronger versions of the rhetoric of inquiry and science (see especially Nelson et al. 1987, Simons 1989, Gross 1996, Taylor 1996, Harris 1997, Locke 2002) blended with Billig's (1991, 1996) rhetorical psychology. 'Rhetorics' articulate versions of reality (Potter 1996) that represent it in specific ways from a range of possible ways; which version is the 'right' one is analytically undecidable *with certainty* (see also Barnes and Bloor 1982), because the activity of making a version accountably warrantable simply adds to the versions – including the version that represents science as rhetorical (Locke 1999b). We are then doomed to representation, a key point of diversion from Miller and McHoul that requires more detailed discussion.

Representational and Ideological Matters

Like Miller and McHoul, one of the strengths I see in cultural studies is its break from the conceptual apparatus of 'mass' culture with all the attendant worries over media 'effects' discussed above. This was initially instigated by the Birmingham Centre for Contemporary Cultural Studies (CCCS) seeking to replace negative images of working class subculture(s) with a more positive view from a broadly marxian problematic as 'resistances' to the dominant culture of capitalism (Hall and Jefferson 1976, Willis 1977, Hebdige 1979, Hall et al. 1980). Their focus on media consumption contributed to the notion of the 'active audience' (Morley 1980, Moores 1993), culminating in an increasingly celebratory view of popular culture (for example, Fiske 1998). However, this only served to expose more fully the abiding tension between seeing culture as a product of powerful groups' impositional and restrictive practices, as opposed to the activity of ordinary people using these as resources to make their own ways of life (expressed in the sharp exchange between Garnham 1998 and Grossberg 1998; see also McGuigan 1992, Ferguson and Golding 1997, Stevenson 2002). These debates having 'worn themselves out' (Curran and Morley 2006: 1), efforts at broader, more 'neutral' conceptions of culture have been advanced, such as 'signifying practices' (Barker 2008: 7) and 'assemblage[s] of imaginings and meanings' (Lewis 2008: 18). Such understandings have enabled closer alignment with the sociology of knowledge (McCarthy 1996, Chaney 2002), bringing us back to Berger and Luckmann's seminal social constructionist concern with 'everything that passes for knowledge in society' (see also Burr 2003, Gergen 2009).

In these conceptions, meaning is an active process of sense-making centrally involving language as the exemplar of signifiying practices, but ultimately incorporating all aspects and forms of human activity for their symbolic-representational and meaning-constitutive character. In line with pragmatic and phenomenological traditions in social theory (Atkinson and Housley 2003), meaning is seen as continuously being generated from available symbolic resources through interactive negotiation, in which individual identities are forged and re-forged in the course of ongoing encounters that might incorporate any relevant material from sedimented stocks of knowledge and objects to hand. This might – and arguably increasingly and inevitably does – include knowledge and objects derived from 'technoscience', which in becoming more central to the established institutional order of modernity has become part and parcel of everybody's everyday lives (Bell 2006, Michael 2006).

Such conceptions of heterogeneous and hybrid identities – or 'ethno-epistemic assemblages' (Irwin and Michael 2003) – filled with images of flux and flow in an ongoing stream of social life, seem close to the ethnomethodological interest in the ongoing practical reasoning of 'immortal ordinary society' (Garfinkel 1991) from which there is 'no time out' and a world – or at least, worldview – away from CCCS' concern with clearly defined ideological 'subject-positionings'. Nonetheless, in cultural studies a point of connection remains through the focus on 'representation' in which, as Miller and McHoul (1998: ix) define it, 'everyday cultural objects stand on behalf of ... wider social forces' such as 'sexuality' or 'race' (recent examples are Jones forthcoming, Dittmer forthcoming). For Miller and McHoul (1998: 21), however, this shows an over-concern with meaning, especially ideology which is 'always thought of as substantive knowledge; something like how an everyday person thinks of, or relates to, his or her "real" situation'. This 'reality' is disclosed by the cultural analyst who claims greater insight into the dominant ideology, but for Miller and McHoul (1998: ix) this is both too 'speculative' and too concerned with the 'spectacular', focusing on the representational practices and meanings of celebrities and youth subcultures to the neglect of the mundane social world. Against this, they call for a focus on how cultural objects are formed within this mundane world through people's sense-making practices or 'ethno-methods'. Ethnomethodology, in their (1998: xi) version,[4] treats 'what folks are doing' as 'everday events as they are, in their own right'. In particular, it is interested in how people 'select... from the practices available within the rules of a situationally specific discourse' (Miller and McHoul 1998: 2) in order to go about doing what they are doing within that situation. This is inherently reflexive: 'how people make things happen ... is identical with how they make those things intelligible' (Miller and McHoul 1998: 19). Thus descriptions and accounts are analyzed not for what they 'represent' – as instanciations of 'ideological' relations taken to be structuring impositions on ordinary life – but for what they *do* in the immediate interactional context in terms of the sense they make of it and thence how they constitute it as such.

An important element in Miller and McHoul's discussion is Membership Categorization Analysis (MCA), originally instigated by Harvey Sacks as the study of the way people use categorizations of themselves and others as part of their ordinary understanding of the social world. As is more fully discussed in Chapter 9, Sacks (1995a: 113-125) viewed categories as 'inference-making machines' used to deduce things about the actions of individuals to whom a category is applied and to apply categories to explain observed actions. His often cited illustration (Sacks 1974, Silverman 1998, Baker 2000, Lepper 2000) is the opening line of a story told by a child: 'The baby cried. The mommy picked it up.' Sacks argued that our ordinary inference is that the 'mommy' is the mother of the baby, partly because picking up babies is what mothers do and because a mother will ordinarily do so when her baby cries. As he (1995a: 241) put it there are 'category-bound activities'

4 Ethnomethodology self-exemplifies in that different versions of it have been advanced and advocated since its inception not least by Garfinkel himself (1967, 1974, 1991, 2002; see also Pollner 1991, Rawls 2002, 2009, Cuff, Sharrock and Francis 2006: 143-171, Lynch 2009).

(CBA), actions ordinarily linked to categories that enable such inference-making. Thus, Miller and McHoul (1998: 124) argue 'inferences *go backwards* from categorial explanations to describe events'. The mother's action is explained by virtue of the category used to describe her as such. No assumptions need be made about this particular mother's internal mental state, her motivations, intentions, attitudes or personality, as the category alone is sufficient to explain the action rather than anything about the particular person within. This is how membership categories are employed ordinarily as 'machines' to make inferences about actions, without need to refer to the 'ghost in the machine', as it were.

It can readily be seen how this might be applied to a term like 'public understanding of science', a phrase containing a membership category, 'public' and a CBA, 'science', with an implicit contrast between them (see also Cooter and Pumfrey 1994: 253-54). Employing the 'inference-making machine', we would infer that 'science' is an activity associated specifically with scientists: science is what (only) scientists do. 'Public', on the other hand, is a category into which everybody else falls, their membership being defined by the negative quality of not being a scientist. Almost by definition, therefore, the inference-making machine seems to lead to the conclusion that the public do not and perhaps cannot understand science. Given this, some version of the deficit model seems inevitable. Moreover, there is a moral order to this (Jayyusi 1984): the public qua 'non-scientists' *should not* claim to be scientists or do science, because science is an activity category-bound to scientists. For them to presume to do so constitutes a normative breach (Garfinkel 1967: 35-75) likely to provoke strong negative moral sanction, as is clearly apparent in the notion of 'anti-science' and the overt moral language sometimes employed to describe such groups as creationists (Kitcher 1982; see also Lessl 1988, Cremo 1998).

On the other hand, in the light of the earlier discussion, this brief application might also raise some doubt as to whether these are necessarily the inferences we do actually make. This points to some problems with MCA as Miller and McHoul use it, which come out from one of their specific applications. They (1998: 125) contrast MCA with standard psychological/ psychotherapeutic analysis, which turns 'normal' category-based inferences into internal states of individuals that are then used to explain actions. So, in the above example, the 'mommy' picking up the baby might be viewed as a measure of her 'motherly-love', say, which then might be made a variable to be measured and tested in laboratory experiments (to determine if it is 'natural' to women), or perhaps employed as a warrant for therapy (if, say, a mother did not pick up her crying baby). Miller and McHoul (1998: 125-126) use the example of 'intuition' in what they call 'pop psychology', citing a case of a pet's apparent foreknowledge of when its owners were about to arrive home as being due to 'a finely-tuned sense of perception', to which 'expert' testimony was added to the effect that people in western societies ignore their 'sixth sense' and are 'disconnected from their intuition'. Miller and McHoul gloss this as follows:

> some individuals have a particular inner capacity that resides in the "subconscious" and is a psychic corollary of the bodily senses. Almost anything that is explicable by normal inference or is generated by probability (such as getting a call from someone you were just thinking of) can then have "intuition" super-added to it.

Against this, they present an 'inference-making machine' version:

> The ordinary inference is visible: parents return from work at more or less regular times; if they have animals, that is when the animals get food and affection. Animals are not without a sense of time, if only the regular timing of food and affection. *Ergo*, the cat waits for the sound of the returning car.

Thus they conclude: 'Intuition is no more than the inference-making machine used by every member of "rational" society. The psy-help account of it … turns it into something extraordinary.'

However, there are some problems here which have implications for PMS. To get to these, there is an issue to address first about Miller and McHoul's critique of representations and the attribution of meaning as stuff inside people's heads. In support of their rejection of this in favour of the study of members' methods of sense-making, they (1998: 134) quote Sacks (1995a: 468) as follows:

> one kind of problem a culture faces is getting its known things kept alive. A basic thing it uses is people's heads. Where people's heads are not just to be repositories for known things, but they have to be repositories that are appropriately tapped so that those known things get passed to others.

For Miller and McHoul, it is how this 'tapping' works that should be the focus of study, the methods used to pass knowledge on rather than the 'known things' themselves. But we can surely use this same quotation to make a case for studying these 'known things' too; arguably, these are of more importance since there would be little point in having methods of tapping if there was nothing there to be tapped!

One response to this might be to point to reflexivity and argue that the 'what' and the 'how' are identical: 'To recognize *what* is said *means* to recognize how a person is speaking' (Garfinkel 1967: 30). Thus, the activity of tapping things constitutes those things as what they are – for example, referring to rules as a method of 'making out' that activities are in accord with the rules and thereby constituting those activities as 'rule-following' (Zimmerman 1971: 233), or constituting something as a 'discovery' by virtue of the activities involved in 'doing discovering' (Garfinkel, Lynch and Livingstone 1981; see also the exchange between Lynch 1992 and Bloor 1992). The trouble here, however, is that such accounts are examples of the phenomenon they supposedly describe, because in describing they re-formulate ordinary reasoning in terms other than its own;

thus, the 'how' (the re-formulated version) becomes the 'what'. In Garfinkel's (1967: vii) terms, then, these accounts are 'arguing correctives' and so are 'done as ironies', because making distinctions between the 'what' and the 'how' is itself an ethno-method, something folks do – for example, by claiming that their activities were determined by the rules or that it was their intuition that told them who was about to phone. Such claims constitute some of the 'known things' of modern western cultures and they are representations in the specific sense given by Miller and McHoul: they treat actions as standing on behalf of wider forces which are taken to have determined them (within which 'social' forces are one sub-group – see Chapter 8 and Locke 1999a, 2009c). As such, they are legitimate objects of sociological study and analysis, not merely as methods of accounting, but as representations of reality and indeed as ideological.

'Ideology', however, does not have to be seen as a mask disguising a hidden reality into which the analyst has special insight. Rather, it can be viewed as the available rhetorical resources of representation that provide contrasting ways of defining the situation. Such contrasts pose us dilemmas (Billig et al. 1988) and lines of tension (Johnson, Dandeker and Ashworth 1984) that provide basic resources for thinking and arguing (Billig 1996), that is things both inside *and* outside our heads – and just which ones are located where is itself a matter for thinking about and arguing over. As such, these 'things' are interactionally derived ('tapped') as both 'seeds and flowers' of argument (Bacon quoted in Billig 1996: 235) and continuously fed back into emerging interactional contexts, flourishing further as they go. One set of seeds is the commonsense contrast contained in the sayings, 'the pen is mightier than the sword' and 'sticks and stones may break my bones but words can never hurt me'. The former has flowered into a philosophical argument to the effect that words do things; but the latter continues to flourish in response in the form of critical realism (Delanty and Strydom 2003, Blaikie 2007; see also the exchange between Billig 2008a, 2008b and Fairclough 2008a, 2008b). Ethnomethodology, in other words, is positioned within this ideological dilemma, which is as much a matter of ordinary commonsense reasoning as it is a concern of academics. Through arguing over, with and through such dilemmas, reality is continuously being re-defined by mobilizing available resources of representation that are reworked and newly applied in and as the activity of arguing. Thus, established rhetorical 'stopping points' (Prelli 1989a) and 'commonplaces' (Myers 1985, Myers and Macnaghten 1998) are reiterated, even as new arguments and new rhetorics are invented (Locke 2002).

The 'Spectacular' and the 'Ordinary'

Miller and McHoul's own text exemplifies this as it provides examples of the very features they critique about cultural studies, specifically in that they mobilize representations that are speculative and rely on a contrast structure between the 'spectacular' and 'everyday life'. Thus, with respect to the above argument about 'intuition' – a representation of reality – Miller and McHoul do not do without speculation about internal states, but merely shift this from people to pets. Specifically, the 'ordinary inferences' they draw on, that animals 'are not without a sense of time' and engage in activity called 'waiting', are anthropomorphic and as such no less speculative than the idea of a 'sixth sense'. They represent the actions of animals as standing on behalf of something else. To describe an animal as 'waiting' is to take its actions as signifying a form of internal state, the more so if it is said to be 'waiting for food and affection'. This is to attribute specific motivation and intention to the animal; it is simply a different type of motivation and intention than 'intuition'. Miller and McHoul's representation of this reality is, then, rhetorical as it is designed to do argumentative work to persuade us against the reality of intuition as an internal psychological state and towards an essentially materialist worldview in which such phenomena as 'psychic senses' do not exist. From my point of view, however, I take it that whether or not intuition and psychic senses do actually exist is properly a matter for members, as also is whether to account for reality in materialist or idealist terms.[5]

This critique can be broadened to other aspects of MCA and to EMICS in general. That MCA is speculative is apparent from the examples used above of the child's story and PUS. Is it the case that we necessarily assume that the 'mommy' is the specific mother of the 'baby'? Just as in the case of the multiple possible interpretations of 'PUS', what is overlooked is the possibility of alternative readings – the point cultural studies makes in its stress on polysemy and the role of the reader in constituting the meaning of texts (Bennett and Woollacott 1987, Barker and Brooks 1998). This resonates with Schegloff's (1995: xlii) argument that MCA is 'promiscuous', relying too much on the interpretative work of the analyst (further discussed in Chapter 9). The strong view here is that the analyst is always at the end of the chain of analytical interpretation; there is, even for analysts, 'no time out' from the activity of sense-making. In employing the technical analytical vocabulary of EMICS to re-describe and thereby re-formulate people's actions in their own terms (Billig 1999a), Miller and McHoul are being no less 'speculative' than the cultural studies they criticize.

Here, again, the question of reflexivity comes in. In Miller and McHoul's (1998: 22) terms, reflexivity considers 'how certain social activities are carried out by talking in a particular way' and what 'purposes' the talk serves – a question then we might ask of their own discourse: what social activities are carried out by Miller and McHoul's text? Obviously enough, their text was written in their membership category as academics and in a range of ways, not least its technical vocabulary, it accomplishes 'being academic'. It does so in both general and specific ways, through the particular literatures cited that help legitimize their academic status as certain categories of 'expert', and warrant the specific arguments and claims they

5 And, yes of course, I am also a member of a modern western society. Accordingly, I have my views about the reality of such phenomena and my own philosophical commitments (or conundrums), but for the purposes of analysis at least I try to remain agnostic.

make (Gilbert 1977). Citations are visible representations of this status as signs of academic social positioning and the purported authority it carries; thus, they are ways of making 'being academic' happen even as they make it intelligible. They are then rhetorical representations intended to persuade.

Am I then saying that this is the 'real', 'hidden' or 'dominant' meaning behind Miller and McHoul's text? No – citations are hardly hidden! More importantly, the view above might be contested (not least by Miller and McHoul) precisely because it is a *representation* of their textual activity and as such one interpretation amongst possible others. It is ordinarily the case that people present each other with different, often competing representations of reality (Pollner 1974, 1975, 1987, Edwards and Potter 1992, Potter 1996) and amongst the 'known things' they may include is stuff drawn from science. For example, as Miller and McHoul rightly point out, people routinely ascribe mental states to animals including such things as 'sixth sense' and these may be treated as objects of speculative analysis and inference – and not just by 'pop'(?) psychologists, but also by 'pop'(?) biologists like Rupert Sheldrake (1994). I interject the question marks to encourage a little speculation about what Miller and McHoul are doing in describing some psychology as 'pop'. Might they be downgrading it in relation to their, presumably, 'non-pop' ('high-brow' then?) academic work? If so, it looks like we have yet another version of the good old, bad old deficit model.

The irony here is not just done, it is devastating, because for all their arguments against the overly 'speculative' nature of cultural studies, Miller and McHoul's own argument shares a similar rhetorical form. Not only does it rely on speculations about internal mental states, but it also imposes an analytical representation on 'what folks are doing' in such a fashion that those activities are downgraded. With this, Miller and McHoul engage in the same form of reasoning as professional sociology that Garfinkel (1967) was so scathing about: they treat people like 'judgmental dopes'. In effect, they are arguing that people misunderstand their own reasoning processes getting them backwards and as a result mistake a make-believe fantasy (a speculation about an internal mental state called 'intuition') for a real phenomenon. They then naively fall victim to con merchants in the 'pop-psych' business who make claim to an expertize that can only be fake. This is an argument in the classic rhetorical form of 'debunkers' regarding the existence of 'paranormal' phenomena (Hess 1993; see also Collins and Pinch 1979, 1982, and Chapter 7 below). Miller and McHoul are claiming to have definitive knowledge that the phenomenon of 'intuition' does not exist, because it is a mistaken attribution of category-bound activity. The P.T. Barnum postulate loudly resounds!

Miller and McHoul make this argument in part against the focus of cultural analysis on the 'spectacular', but here again they rely on the same rhetoric in the form of a contrast structure with the 'ordinary'. A number of cultural analysts have made similar critiques (Hermes 1995, Billig 1997), while others have looked in detail at aspects of the everyday in its 'historical particularity' (for example, Jenkins 1992). These texts employ the same contrast structure to position their accounts as closer to the reality of 'everyday life' than other forms of cultural analysis. So, Jenkins is strongly critical of what he sees as the tendency in both the media and academic literature to represent fans of television shows like *Star Trek* as an extraordinary 'other' (see also Jenson 1992) – although, significantly, fans themselves sometimes accentuate the extraordinary nature of their activity in contrast to 'mundanes' (Jenkins 1992: 88) and their own 'everyday' lives. Hermes, on the other hand, criticizes the focus on such 'spectacular' fan-groups arguing for the need to study 'mundane' cultural activities like magazine reading. For this, she draws on Potter and Wetherell's (1987) notion of 'interpretative repertoires' (see also Gilbert and Mulkay 1984, Edley 1993, Locke 1994, 1999a), a form of analysis that has itself been criticized (Wooffitt 1992, 2005) in terms that echo Miller and McHoul's argument that cultural studies is insufficiently 'ordinary'.

Thus, the 'ordinary'/ 'spectacular' contrast structure is employed by each of these analysts to represent their quite different analyses as closer to the 'everyday'. Like other contrast structures (Smith 1978), this establishes a normative baseline, one in which 'ordinariness' is used to make other activities appear extreme and unusual. There is a curious inversion here of the traditional narrative form of scientific texts presented as journeys into the unknown (Brown 1994), as these authors are making the peculiar claim that what is unknown to academics is the very mundane reality they also claim we all share. Nonetheless, the standard rhetoric of discovery accounts remains; thus, the appeal to the 'ordinary' lends a form of authenticity, just as in fan studies it has become important to declare oneself a fan to demonstrate credibility (Hills 2002; see also Delamont 2009). This, then, is a presentation of self designed to convey a particular ethos (Edmondson 1984) as a means to persuade readers that the account given is 'true' to the reality described, the way the social world is (Mulkay 1985). In the terms of conversation analysis, it is a way of 'doing being ordinary' (Sacks 1984) without actually being 'ordinary' at all because one is actually being an academic (who as we all know are most extraordinary people!).

However, doing 'being ordinary' is not just confined to academics, but is also done by ordinary people like the fans who accentuate their extraordinariness and by people who claim to have had paranormal experiences (Wooffitt 1992). Thus, the 'ordinary'/ 'extraordinary' contrast is itself a feature of ordinary sense-making about the world (Chaney 2002), commonly where this involves something of a technoscientific nature like science fiction or the parnormal. As such, it is a *constitutive ideological dilemma* of mundane reasoning about science in modern culture – and this brings me back to rationalization.

Science fiction is briefly addressed by Miller and McHoul (1998: 149) when they apply MCA to the cult-tv show *The X-Files*, stating that 'one of [its] unique qualities ... is that it works by building up a puzzle and ... offering a solution in the course of the narrative; or else ... leaving the reader or viewer with a conundrum to solve in the form of a reading or interpretation'. This is often done by presenting 'unknown' agents such as alien or monstrous entities, which cannot readily be membership categorized and whose activities are not easily inferred. Thus, they (1998: 150) assert that this 'is quite specific to science fiction and horror genres', a claim that strikes me as quite unjustifiable since other genres, such as murder-mystery and spy fiction, often work in a similar way. In these cases, the 'unknowns' may usually be people, but they are made puzzling as agents by virtue of the fact they may be murderers or spies. Much murder fiction turns on the fact that someone who is thought to be known and category-bound in their activities turns out not to be, such as the proverbial butler.

What most disappoints about this is that Miller and McHoul overlook the fact that the definition of science fiction has been an issue for writers and readers since Hugo Gernsback first sought to define 'scientifiction' as a recognizable form of literature in the 1920s (Parrinder 1980, Aldiss 1986, Bainbridge 1986, Lambourne, Shallis and Shortland 1990, Clute and Grant 1997, Vieth 2001, Erickson 2005). Once again, they simply impose their analytical interpretation. If they had attended to ordinary members' interpretations (not to say representations) of science fiction, one of the things they might have been drawn to consider is that whether or not *The X-Files* is an example of 'science fiction' is by no means straightforward. In some definitions of 'hard sf' (as opposed to 'soft sci-fi'), the programme is insufficiently grounded in accepted scientific 'fact', containing too much 'unscientific' speculative material, often presented on-screen as ostensibly 'real', about purportedly 'paranormal' phenomena such as alien visitations to earth (Mellor 2003: 532, n.7). We are back with the Drake equation! Other views of sf (and/or sci-fi) might suggest otherwise – just as for some (like Clute and Nicholls 1993) paranormal beings like superheroes are not science fiction but 'science fantasy', whilst for others they are science fiction (see Chapter 6) and for still others whether they are science fiction or not depends on which particular powers we happen to be talking about (Gresh and Weinberg 2002).

The point here is that writers and readers of science fiction, whether viewed as 'spectacular' fans or as 'ordinary' members, take as a matter of *their* concern where the boundary line between such categories as 'fact', 'fiction' and 'fantasy' is to be drawn. Within this, the *meaning* of science is of central significance as a resource of legitimation, but also of argumentation with which to think through and about 'reality'. The meaning of science becomes a site of cultural struggle both as a general rhetorical resource, a suasive device to defend one view and oppose others, and also as a resource of specific materials – 'facts', theories, methods, perspectives, arguments etc. – with which to construct and articulate social divisions and identities (or membership categories). Amongst these resources are those of the 'spectacular' and the 'ordinary', generating amongst other things rhetorics of enchantment (see Part II) and mundane mystery (see Part III). These are alternative characterizations of what science is and what it does, resources of argument over its nature and meaning.

Expressed in Weberian terms, science is *both* disenchanted *and* enchanted, and it *both* disenchants *and* enchants (McPhillips 2006). To make this claim, however, is to contradict the standard view of rationalization and therefore, evidently, to contradict Weber (Tiryakian 1992). I do not see it this way; rather, I see the development of both disenchanted and enchanted versions of science as a continuing working out of the logics of rationalization as Weber typified them, albeit through a paradox of consequences he did not perhaps consider. Thus, the case for this requires some re-crafting of his thesis. It is partly this use of Weber that distinguishes rhetorical sociology from some other recent approaches to public science (Irwin and Michael 2003, Erickson 2005, Bell 2006, Broks 2006, Michael 2006). In addition, the imagery of 'crafting' is important. I use this deliberately to invoke the sense of the artisanal and more so the *bricoleur*, which for all its *passé* structuralist connotations flatters my vanity to think of myself as an intellectual craftsman (Mills 1999) and sits well with the sense of what I am doing as tinkering around with everyday, speculative and spectacular social constructions of science. In addition, an older meaning of 'craft' is significantly cued: the enchanted craft of the witch. In re-crafting rationalization, I am doing both of these things: reconstructing Weber's conception (or, rather, our conception of his conception), and doing so through and by reference to enchantment. What that means, I now try to explain.

Chapter 2
Intellectualist Rationalization, or What Does it Mean to Live in a Scientific Society?

In Chapter 1, a view of PUS as concerned with public meanings of science was outlined, coupled with a strong claim that this is properly a matter of concern to sociologists. Fundamentally, PMS asks the question, what does it mean to live in a scientific society? This question concerned the founders of sociology, but is perhaps most explicitly put by Max Weber. At any rate his reply has been widely taken up in contemporary sociology in respect of two main features: a view of the prevalence of instrumental rationality with an associated orientation of mind that Weber called 'disenchanted'; and an associated process of secularization in which religious and 'magical' outlooks in general progressively lose their influence over the organization of practical social activity and the attitude of mind of individual actors (the classic statement being Wilson 1966). Thus, in what I call the standard view of rationalization, disenchantment is seen as the principal outcome of 'intellectualist rationalization', which Weber (1948b: 139) linked directly to 'science and ... scientifically oriented technology' and meant that 'there are no mysterious incalculable forces that come into play, but rather that one can, in principle, master all things by calculation.' For Weber (1976: 13-31), this is to be understood as part of a wider process of rationalization marking the historical uniqueness of the modern west, in which what he took to be the formally rational characteristics of different 'value spheres' of society (notably the economy, polity and intellectual/ spiritual) develop to their progressively fullest realization. From this has developed a prevailing instrumental attitude of mind marked by calculative, means-ends reasoning concerned with achieving a given goal by the most logical means possible without regard to other, specifically moral considerations.

There are a number of points to be made about this thesis and its conventional uptake: first it is essentially a claim about PMS; second this claim can be challenged in the terms of the thesis itself; third doing so provokes a challenge to standard theoretical uses of the thesis, but also clarifies ambiguities within them. The core of the argument is that the standard view does not do full justice to Weber's analytical framework, which actually allows for a wider range of potential outcomes of intellectualist rationalization than he himself accentuated. His view of disenchantment rests on only *one* version of science and its public meaning that may be derived from this analytical framework, a version he emphasized because of his concern with its moral implications for the practice of the scientist (Lassman and Velody 1989) and which he derived from the ideal-typical analytical model built around his notion of *formal rationality*. As such, it overlooks – or at least downgrades – alternative public meanings of science that are possible *within* the framework of this ideal-typical model with respect to *substantively rational*, and both *formally* and *substantively 'irrational'* meanings of science (collectively, 'informal' meanings). It needs to be emphasized that the terms 'rational' and 'irrational' are intended in a relative fashion in relation to the perspective taken from each value sphere, as will become clear from the ensuing discussion. The notion of 'formal rationality', however, is a little more troublesome as will be seen.

My re-crafting of Weber, then, is made in the spirit of Weber and is not intended as a critique so much as an attempt to work through the implications of his analytical model in a fuller way than he himself seems to have done, specifically with respect to science and its public meaning. The alternative possibilities this gives rise to are more fully elaborated in Chapter 3, but first the standard view needs to be set out to justify the claim that this can properly be seen as about PMS and to clarify the alternatives. Accordingly, the discussion focuses only on those aspects of Weber's work most relevant to this purpose. In addition, to help show the full significance of my re-crafting, I give brief attention to two influential uptakes of the standard view in recent social theory: Lyotard's (1984) 'condition of postmodernity' and Beck's (1992) 'risk society'. These theories are of particular interest because although they relate significantly different narratives about the direction of development of contemporary society, they do so against a background of shared assumptions about science and its social significance that owe a good deal to the Weber model. Moreover, making these assumptions explicit shows that they articulate two dilemmas within PMS regarding the nature of science and its social impact. This has wider implications for sociological understandings of a range of contemporary phenomena, including some linked to secularization that I turn to in Parts II and III.

The Ideal-typical Weber Thesis: Toward a Disenchanted Society

Weber's rationalization thesis has been widely discussed and it is not my intention to provide an overview of this literature (recent accounts include Gane 2004, Poggi 2006; see also Tiryakian 1992, Lee and Ackerman 2002, Partridge 2004). Instead, I am going to present an account based largely on my own reading of English translations of Weber, especially those on religion (Weber 1948a, 1948c, 1951, 1952, 1958, 1965, 1976). This aims to bring out two interconnected points. First, it is possible to read his thesis as informed by an ideal-typical view of rationality marked by specific formal characteristics that accentuate an instrumental mental orientation (*Zweckrationalität*). It is on this basis that Weber views modernity as an unique historical epoch in which it has so happened that the formal rationalities of several value spheres (notably, the political, economic, intellectual and legal[1]) have begun to mutually reinforce, such that he considered the probability of maintenance of any value-rational mental orientation other than the instrumental to be increasingly unlikely and for the scientist in particular to be morally culpable. This latter point makes clear that he nonetheless considered instrumental action to be a form of value orientation with an associated 'ethic of conduct' and this opens up one way to begin to develop an alternative view of PMS from the thesis.

Secondly, in his detailed comparative historical analyses of non-western societies, Weber tries to show that formal rationality has been restricted in manifesting its ideal-typical form of development both by the specific direction of rationalization ('world-rejecting') and by a variety of substantive rationalities arising from the interests of specific social groups, such that in one way or another the logics of different value spheres worked against each other rather than mutually reinforcing. As a consequence, although intellectualist rationalization did lead to other types of disenchanted outlook, these did not culminate in the purified instrumentalism he emphasizes in western science. But even in his account of the west – notably in his view of the relationship between bureaucratization and capitalism – he points to the continuing presence of substantive interests restricting the logic of formal rationalization whatever his assessment of the likely future (Eldridge 1971). This opens up another way to begin to develop an alternative view of PMS.

In both cases, it is not the adequacy of the ideal-type as an analytical model that is of concern, but how Weber chose to use it as a basis on which to extrapolate future possibilities. Here, it needs to be stressed that an ideal-type is precisely *not* an empirically occurring type. Actual social action only approximates to a pure type (Weber 1968: 26). Accordingly, Weber's view of the implications of science for the modern outlook needs to be understood not as empirical description, but as an *argument* and thus as a rhetorical characterization that deliberately accentuated *one* version and *one* possibility over what are in fact multiple versions and possibilities. This is apparent from the rhetorical context documented in essays written in reply to Weber's thesis by his contemporaries (collected in Lassman and Velody 1989; see also Harrington 1996.) Thus, despite Weber's strong statement to this effect, disenchantment is *not* a necessary implication of the rise of science, nor necessarily widespread and *nor is there any reason to assume it has ever been* within modern culture. One implication of this is that, contrary to a growing body of literature (Griffin 1988, Tiryakian 1992, Lee and Ackerman 2002, Lee 2003, 2007, 2008, 2009, Saler 2003, Partridge 2004, Foltz 2006, Letcher 2006, Possamai 2006), there is no need to write of *re*-enchantment, because the west has never been without enchanted outlooks, some of which have been and are drawn from science (Harrington 1996, and see Chapter 4 and Part II below).

1 Weber scholars dispute the treatment of law as a separate value sphere (Gane 2004: n. 1, 161-162). I separate it in order to make a comparison between Weber's model of rationality in law and that in science (see Chapters 3 and 4).

It should also be made clear that my own account of Weber's rationalization thesis is itself an ideal-type, with no greater status than an account constructed for specific analytical – and rhetorical – purposes. Those purposes are to raise doubts about the adequacy of some views of contemporary society in respect of their assumptions about PMS and elaborate an alternative. Like any ideal-type, mine is grounded in empirical referents but accentuates certain features for analytical purposes, specifically intellectualist rationalization to the relative neglect of economic and bureaucratic rationalization. I do not deny the importance of these to Weber's overall thesis – arguably, the latter is far more fundamental. Nonetheless, my interest is in PMS and Weber's view of this can be discussed separately as, while he (1958: 151) held that 'practical rationalism [is] the intrinsic attitude of bureaucracy to life', this is not the same as disenchantment itself.

My empirical referents are chiefly Weber's writings on religion (as Parsons [1968] argued, for the theorist, theories are empirical materials). Given the enormous scope of Weber's oeuvre in which the directing vision can easily be lost to sight (Hennis 1983, Gane 2004: 4-7), limiting the focus in this way helps to keep the argument on track. Moreover, it is in these texts that he develops his view of intellectualist rationalization, fundamental to which is his 'philosophical anthropology' (Poggi 2006: 21). This is about *meaning*; as Poggi states, 'human beings are on the one hand compelled, but on the other hand enabled, to locate themselves in the reality within which they exist ... and to act within that reality, on the basis of the *meanings* they attribute to it.' In Weber's view, such meanings are essentially arbitrary because they are not given by reality itself, but they must nontheless be lived as though they *are* real. Poggi (2006: 22) puts this neatly: 'human beings by their nature *inscribe* meanings *on to* reality, but must delude themselves that [they] *read* such meanings *off* reality.'

Whilst this sounds a little too close for comfort to a theory of ideological delusion, it does presage the crucial point about science that in Weber's view an understanding of the facts (what is) can never provide a basis of moral decision (what ought to be) – and the seeming paradox that this is itself presented as a *moral* position fundamental to the meaning of science in modernity. Further, any discomfiture over the delusional implications is offset by recognizing Weber's sense of the essential tragedy of the human condition that comes out more fully in his view of religions as projected solutions to the problem of *suffering* (Shafir 1985). Religious teachings are understood as 'theodicies', providing an explanation for the existence of evil and the means whereby it may be overcome – a route to salvation. Although religions act to legitimize existing distributions of secular power and privilege, historically their doctrines have undergone rationalization to be reformulated in the apparent interests of the oppressed (Weber 1948c: 271-76). From once having been evidence of rejection by the gods, suffering became the mark of salvation to be achieved through the specific soteriological technique prescribed by the belief. So, whilst religions provide their followers a meaning of life that is essentially arbitrary, they do so through a guide to individual behaviour – an ethic of conduct – of profound psychological power and world-changing practical significance. Therefore, even if people are deluded, they work hard to make the delusion of real benefit to themselves in terms of *their* sense of what is important – and where Weber differs from other views of ideology (for example, Eagleton 1991) is that he makes no claim to ultimate judgement as to the validity of such beliefs; nor in his view can or should the (social) scientist. In a sense, we are no less deluded than they are, except it is 'the fate of our times' to be aware of our condition and be powerless to mitigate it.

However, the precise form and direction of practical conduct is dependent on other features of the specific social context. So, although religious beliefs are in one way the foundations of social orders, in another they are subsidiary. This is because the different value spheres each have their own *formal* ethic and different religious beliefs interact with these differently. However, this raises a problem at the core of the rationalization thesis concerning the nature of rationality. Fundamentally, this concerns the problem of the sociology of knowledge: how it is possible to claim universal validity for scientific knowledge when it is also associated with one particular form of society, modernity (Locke 1999a)? Traditionally, sociologists were happy to explain other forms of belief as socially 'caused', but not science, a position challenged by the sociology of scientific knowledge (SSK) (Bloor 1976, Mulkay 1993). However, Weber's rationalization thesis can be seen as an attempt to resolve this puzzle through the distinction he makes between formal and substantive rationalities, with the former tied to the unique features of the modernist outlook. Instrumentalism provides the morally neutral yardstick by which value-rational action (*Wertrationalität*) can be assessed, a view that is both the basis of his sociological methodology and his analytical account of modernity. But this itself turns into a *moral* argument about the ethic of conduct of the scientist, which far from resolving the 'war of the gods' between competing value-rationalities, simply continues it by other means. Instrumentalism is one value amongst others and these continue to inform science, because the features Weber associates with formal rationality are not as definitive as he makes out. Action can always be characterized differently: one person's 'instrumentalism' is another's 'value-judgement' (Schutz 1972: 3-44, Parkin 1982: 17-39). Thus, formal rationality is simply one version, an available rhetorical resource for characterizing the activity of scientists.

Formal Rationality and its Restriction

What then does Weber mean by 'formal rationality'? Some idea comes from the following passage (Weber 1948c: 293-94; see also 1968: 85):

> We have to remind ourselves ... that "rationalism" may mean very different things. It means one thing if we think of the kind of rationalization the systematic thinker performs on the image of the world: an increasing theoretical mastery of reality by means of increasingly precise and abstract concepts. Rationalism means

> another thing if we think of the methodical attainment of a definitely given and practical end by means of an increasingly precise calculation of adequate means. These types of rationalism are very different, in spite of the fact that ultimately they belong inseparately together ... "Rational" may also mean "systematic arrangement" ... In general, all kinds of practical ethics that are systematically and unambiguously oriented to fixed goals of salvation are "rational", partly in the same sense as formal method is rational, and partly in the sense that they distinguish between "valid" norms and what is empirically given.

This provides three senses of 'rationality': 'theoretical mastery' through conceptual abstraction and analysis; 'precise calculation' of means; and 'systematic arrangement' in respect of an ethic of conduct directed at a specific goal of salvation. However, only the first two are strictly examples of 'formal method', which the third employs but towards a goal that cannot itself be considered rational. Similarly, Weber (1949) distinguishes 'technically correct' action from action that may be consciously planned but does not conform to a strictly logical consideration of the link between means and ends.

Two critical points come from this. First, Weber works with a 'formal' sense of 'rationality' that is treated as transcending social context at least for the purposes of analysis (ideal-typically). This is used as the yardstick to develop the account of the specific form taken by intellectualist rationalization in the west and its affinities with the formal rationality of other value spheres. In effect, the formal rationality of all value spheres is treated as essentially the same, involving theoretical mastery and calculation of means in the interests of technical correctness. This is, however, an extrapolation from actual action, which invariably involves substantive rationalities not essential to the specific sphere of action but drawn from other spheres, as well as both formal and substantive 'irrationalities' (see Chapter 3).

Second, in the ideal-type of formal rationality, Weber effectively aligns 'theoretical mastery' with the 'calculation of means', which although he says are 'very different', he treats as 'inseparately together'. This is central to his understanding of western intellectualist rationalization and relates to his view of the role played by the charismatic in historical development. 'Charisma' refers to 'an extraordinary quality of a person, regardless of whether this quality is actual, alleged or presumed' (Weber 1948c: 295). It is possessed by all religious 'virtuosi' and is used to characterize the 'magical' and 'ecstatic' qualities attributed to the world by believers, in contrast both to traditional forms of behaviour resulting from the routinization and habituation of such beliefs, and the 'legal-rational' form of the modern disenchanted outlook. Charismatic individuals lead projected social revolutions, typically by identifying through theoretical analysis logical inconsistencies in the established form of belief that underwrites the traditional social order; hence, magic becomes rationalized (Eldridge 1971: 57). Historically, the outcome of such rationalization has been progressive universalization and impersonalization of beliefs – the gods of the tribe become displaced by the single, universal God, who in turn becomes displaced by abstract 'natural' forces – and increasing 'systematic arrangement' in the logical connection of means to the given end of salvation. In the west, this has culminated in a dominating calculative logic, such that theoretical abstraction has become interconnected with the instrumental pursuit of ends to the point where this threatens to become an end in itself.

In non-western societies, this alignment did not occur for what appear to be three different reasons: the inherent formal rational logic of the particular direction of mental orientation taken by the 'life of the spirit', towards 'other-worldliness'; limits on this imposed by a variety of substantive rationalities associated with other value spheres; and most pertinently, the separation of theoretical mastery from calculative means. These can be briefly illustrated respectively, by the examples of Hinayana Buddhism, Mahayana Buddhism and, in some ways the most interesting case, Confucianism. Weber characterizes the direction of intellectualist rationalization as having moved in two broad ways: 'this-worldly' and 'other-worldly', and, concomitantly, religion has been either 'world-transforming' or 'world-rejecting'. In his work on 'world religions' (Weber 1948a, 1948c, 1951, 1952, 1958, 1965, 1976), the former refers to Judaism, Islam and Christianity; the latter, to Hinduism and Buddhism, with Confucianism somewhat betwixt and between. The accuracy or otherwise of these characterizations is not at issue for the present discussion as the intention is only to set out Weber's model to show how the standard view of rationalization can be re-crafted (for some criticism of Weber's characterizations concerning science, see Turner 1987).

Hinayana Buddhism comes closest to the fully rationalized form of 'world-rejection' in that it completely lacks any soteriological promise of a better world to come and so has no world-transforming capacity or intention (Weber 1958: 204-230). Rather, it teaches that 'this world' is 'cosmic illusion' and that any activity only serves to keep the individual pinned to the eternal wheel of life. The only 'salvation' is to grasp the nothingness and abandon all worldly activity. There is no 'soul' to be saved and even monasteries and other 'communions' are without foundation; the monk seeks a purely personal escape, looking not even to feed him or herself but accepting only unsolicited hand-outs from passers-by. In this respect, Hinayana Buddhism is the effective opposite extreme to the ascetic Protestant sects as the most rational form of 'mystical' world-rejection of no practical use to those, like the Hindu Brahman, who have some vested interest in maintaining a secular position of power and privilege. It is the product of pure intellectual speculation on behalf of a religious virtuoso free from the demands of worldly power. On the other hand, Mahayana Buddhism is a compromise belief with Hinduism and as such, an example of what can happen to an internally consistent religious teaching subject to a priesthood with this-worldly interests (Weber 1958: 244-256). Key doctrinal elements from Hinduism (*karma* and *samsara*) were incorporated to produce a hybrid teaching of massive internal inconsistency that could be employed to legitimize the existing social order in India, the caste system with its dominant priesthood, ensuring the law of *dharma* (caste-specific ethics of conduct) remained barely affected. Thus, in this case, political substantive rationality placed limits on the development of the formally rational logic of world-rejecting belief.

The case of Confucianism (Weber 1951: 142-170), however, is much less straightforward (Molloy 1980) as, strictly speaking, it was neither 'world-rejecting' nor 'other-worldly' – just the opposite. However, neither was it 'world-transforming'. Rather, although its ethic was in some ways comparable to that of the ascetic Protestant sects in being characterized by a pure utilitarianism, this did not take the calculative form of the Puritan work ethic, but was directed towards harmonious behaviour understood as the maintenance of classical order in the running of mundane affairs in which change was considered unfitting to the mandarin or gentleman-official. Thus, whilst almost entirely concerned with this world, compared to Protestantism its vision was myopic, never looking beyond the demands of the moment unless back into the past for guidance in how the affairs of the present should be managed. So, although both generated an ethic of strict sobriety in the face of reality (albeit differently understood), Confucianism abandoned all interest in the extra-mundane in favour of a concern with holding power and privilege within an unified and, compared to Europe, relatively stable empire. As part of this, it tolerated both traditional folk beliefs in animism and ancestor worship, which provided handy means of mass quietization, and the 'ecstatic' spirituality of Taoism once it was seen to pose no immediate threat to 'classical' order (Weber 1951: 205-212).

Thus, Confucianism contained no soteriology as such and provided for Weber (1951: 226-249) a significant comparison to Protestantism in being equally, if not more disenchanted, rejecting all magical explanations of the world. For example, it used the animistic folk belief in the contrary good and evil spirits, Yin and Yang, as a means of philosophical validation for the maintenance of harmony, but rationalized them into impersonal, universalistic principles of balance. Charisma then became associated with the ability to maintain this impersonal balance and the mandarin who could not keep his affairs in order was seen as lacking this quality, thereby threatening the stability of the whole empire and the very boundaries of 'reality'. Further, as charisma was effectively equated with correct behaviour, it was in principle possible for anybody to possess it through ritually correct educational attainment. So, even in this, 'magic' took on a mundane form. Nonetheless, the substantive rationalities of the political and economic concerns of the empire limited the formal rationalization of Confucianism; specifically, although highly rationalized in the form of theoretical abstraction (having a disenchanted outlook) and technical correctness (concerned with correct behaviour), it lacked rationalization in the form of calculative means. Thus, compared to the Puritan, who acted on the world as 'the instrument of God', the 'Confucian … "cultured man" was "not a tool"; that is, in his adjustment to the world and in his self-perfection he was an end unto himself, not a means for any functional end' (Weber 1951: 246; see also Poggi 2006: 87). This points to a disjuncture in Weber's analytical model that will be examined further in Chapter 3.

Similar points pertain to the relative development of science (Schroeder 1995). For Weber (1976), although science exists in all times and places, it is only in the west that it developed systematically. Critical to this were theoretical rationalism in the specific sense of systematic metaphysical speculation through discursive elaboration and extrapolation, and empirical study stimulated by the 'artistic' experiments during the Renaissance (Weber 1948b: 141, 1951: 150-151). Metaphysical speculation is fundamental to the development of alternative conceptions of the world and associated ethics of conduct, and thus important in the rationalizing of prevailing beliefs. Here, he ascribes particular significance to the discovery of the concept in Greece, the 'first great tool' of scientific thought. However, he also comments that Socrates 'was not the only man in the world to discover it', as a similar form of logic was also found in India (see also Weber 1958: 158-162). Moreover, he (1958: 331) states that 'in the area of thought concerning the "significance" of the world and life there is throughout nothing which has not in some form already been conceived in Asia'. Nonetheless, in India the practical impact remained negligible, partly because of the flexibility of the caste system, but ultimately because the extreme devaluation of the 'cosmic illusion' defused any transformative potential. So, too, empirical study: although he (1958: 161-162) states, 'Indian natural science in many areas arrived at a level which Western science had attained about the fourteenth century', again, however, '[i]n the last analysis it was indifferent to the actualities of the world.' In China, contrastively, systematic speculation was hindered by the rigid requirements of the education system directed towards the training of the mandarin in the manners and etiquette of harmonious living (Weber 1951: 119-129). Consequently, the capacity to elaborate alternative beliefs was restricted, limiting the potential for challenges to the existing social order. Meanwhile, Chinese science 'remained sublimated empiricism' (Weber 1951: 151), without theoretical development and so lacking systematic procedures of analysis and experimentation.

It is only, then, in the west that science and intellectual life in general was able to develop a fully rationalized form, taking a 'this-worldly' direction and oriented in a 'world-transforming' way. Crucially, the route taken by the life of the spirit served to further the formal rationality characterizing other value spheres, especially the economic and political, which although tracing their own relatively independent lines of development, have increasingly become mutually reinforcing with the intellectual sphere (Weber 1976: 13-31). Fundamental to this was the unique Judaic conception of the relationship between the god, Yahweh and his worshippers, the tribes of Israel. This was defined through the '*berith*' or 'covenant' (Weber 1952: 75-78, 1965: 246-261), in which Yahweh contracted to deliver the 'promised land' to the loose tribal confederacy that became his 'chosen people', providing they kept his law, especially the injunction to worship 'no other gods before me'. The major consequence of the covenant for Weber (1952: 297-335) was its impact on Hebrew prophecy: whereas in the east, prophecy typically was 'exemplary', the Israelite prophets were 'emissary', bringing the word of god directly in the form of commandments that demanded a change in behaviour from those who broke the covenant. Hence, this was world-transformative and the breakaway sect of Judaism, Christianity, carried this mode of prophecy and its covenantal basis into Western Europe. But the Christian belief in a redeemer, who had taken the burden of human sin upon himself had the two-fold effect of defusing the Jewish inheritance of guilt and dissolving their 'in-group'/'out-group' dualistic morality (Weber 1952: 343-355). Critically then, the tribal Jewish god became in Christianity a universal God (Weber 1965: 115-16), a feature central to its formal rationalization. Thus, the practical ethic of world-transformation that informed ascetic Puritanism, combined with the logic of universalization pursued through Luther's rationalization of the Christian conception of 'brotherly love' and Calvin's radical conception of predestination, set the seal on the fate of the west (Weber 1976).

Modern Disenchantment

Disenchantment stems from this. Fundamental to the Protestant rationalization of Catholicism was the denial of the capacity of priests or other charismatically endowed individuals to intervene for the ordinary person and magically induce salvation. This forced individuals to rely on their own practical action in treating their life as a God-given 'calling', thereby demonstrating their position amongst the 'Elect'. The comparison with Confucianism is telling. For the Confucian, 'salvation' was a meaningless notion; maintaining the balance of nature and hence society in this life was all. In this sense, Protestantism is more 'other-worldly' and to this degree more 'magical'. But its impact on practical conduct was quite different, producing a world where not only was magic denied, but the capacity to judge the significance of life lost its hold on the human spirit. The mandarin, however mortally confined, knew his place in this life and the worth of what he did, but for the Puritan such knowledge came only through the capacity to work without stint in his or her calling – activity that had a most unanticipated and paradoxical consequence, producing a world in which work became all and God was pronounced dead.

Here, then, is how the formal rationalization of the life of the spirit resonates with that of economic and political action. In these spheres, formal rationality is focused on maximizing respectively the returns on investment and control over a given territory. Both, therefore, are impelled ideal-typically by a purely calculative, instrumental logic directed at efficiency and effectiveness of technical means over and above any morally dictated ends. Such ends come from the intellectual sphere. But in this sphere, modern science is also purely instrumental in its pursuit of knowledge of the world, having no interest of its own in moral considerations; so it has no moral boundaries to impose. Economic and political action is then freed from any such restraints, just as the pursuit of knowledge is freed, ideal-typically, from any economic and political interference.

A significant feature of this 'spiritless' modern spirit in Weber's view is not that the ordinary person in their everyday life knows much scientific or technological knowledge – as he (1948b: 139) put it in his lecture on 'Science as a vocation' given in 1918, 'the savage knows incomparably more about his tools'. Rather, it is that we believe in the principle of this-worldly human explanation, so that if we wanted to know, we could find out as 'there are no mysterious incalculable forces that come into play'. This, then, is a claim about PMS: what it means to live in a scientific society is not so much to know what science says nor even necessarily to 'love it' (Turney 1998b), but to believe that the world is knowable in principle through science. But there is good reason to view this claim, even as an ideal-type, as a rhetorical characterization that Weber advances to make a case about the nature of the vocation or ethic of conduct of the scientist. Looked at in this way, it is less about the mental outlook of the ordinary person and more about what he considers to be the proper orienting attitude of the scientist. For scientists to be true to their vocation, they must have the moral fortitude to confront the implications of this activity with integrity, which means they must remain mute on the matter of its worth. The proper scientist confronts history as an eternal path of 'progress', in which her own best efforts are destined to be overshadowed and surpassed. Unlike the Renaissance philosopher, for whom 'science meant the path to true art and that meant … the path to true nature' (Weber 1948b: 142), or even the Protestant, for whom it was the way to God, for today's scientist there is no such hope. Weber is sharply forceful on this: 'Who – aside from certain big children who are indeed found in the natural sciences – still believes that the findings of astronomy, biology, physics, or chemistry could teach us anything about the meaning of the world?' Science has no voice to speak on this; it can comment only on technical correctness and practical know-how undertaken without purpose except for its own sake. In its purified instrumentalism, it stands outside the 'eternal war of the gods', the endless competing valuations informing *Wertrationalität*. Note, however, that the fight does still go on, it is just that the scientist can merely keep the score-card.

It is in this sense that the scientist has a 'universal' basis on which to judge other ways of life – because *Zweckrationalität* is free from partiality with respect to *Wertrationalität*, the only 'pure' comprehension of another's action is possible from the attitude of the former. If the other acts purely instrumentally, then their motivation can be immediately apprehended through the technique of *verstehen*; if not, then the *verstehende* is limited to deducing the possible value-rationality informing the action. The scientist, if true to her vocation, is impartial, but only in this very specific sense. This, then, is Weber's solution to the problem of the sociology of knowledge. It relies crucially, however, on the supposition that a clear distinction can be made between *Zweck* and *Wert*, something called into doubt by Schutz (1972) and shown to be empirically problematic in the context of scientific controversies by, amongst others, Gilbert and Mulkay (1984). Simply put, even amongst scientists, one person's *Zweck* is another's *Wert* and vice versa; thus, the distinction is more deeply analytically problematic than Weber suggests. Part of the reason for this is because, contrary to Weber's moral demand, science remains informed by a variety of competing substantive rationalities or social interests. It follows from this that there is much more to PMS than just disenchantment.

Why not then simply abandon Weber's model of rationalization? I want to argue against this, although we should question the claim that instrumentalism is the prevailing mental orientation either of scientists or the ordinary person. Still, this *does* have a powerful presence in modern culture *as a rhetorical appeal* and in this respect, Weber's ideal-type remains valuable as a description of a formal, publically established, legitimizing discourse. However, we need to recognize that this is only *one* rhetorical characterization of science amongst others, a view for which Weber himself provides some oblique support. For one thing, the fact that he made such an argument itself evidences the presence of competing viewpoints, however dismissive of them he may have been. In addition, he also points to two contrasting tendencies in modernity: whilst he sees the world set on the path to monolithical formal rationalization, the 'iron cage' of 'escape proof' bureaucratization (Giddens 1971: 182), he (Weber 1948b: 143) also notes a resurgence of a form of 'polytheism' amongst the young, turning away from the cold and calculating emptiness of the vocation of science towards the Romanticized emancipation of the 'irrational' (McPhillips 2006), especially in the aesthetic and erotic spheres (Gane 2004).

From this, Weber suggests two further possibilities: in one, through a 'paradox of consequences', the irrational itself may become intellectually rationalized; in the other, there may be a revised 'war of the gods', albeit expressed as 'impersonal forces' rather than personalized entities (Weber 1948b: 147-149). This last development, in Weber's (1948b: 153) morally laden rhetoric, would be a result of 'weakness' by scientists and academics unable or unwilling 'to countenance the stern seriousness of our times' and accept that their responsibility entails a non-judgemental attitude to the issues of the day, avoiding any temptation to turn science into 'the ersatz of armchair prophecy'. But, as shall be seen in Chapter 3, something that might well be considered such 'ersatz prophecy' does indeed characterize a good deal of public science and, I shall argue, for good reason. What it points to is not disenchantment as a widespread public meaning of science, but the significant role of *enchanted* representations as necessary rhetorical resources in the maintenance of institutional science. Thus, whether or not we agree with Weber that such representations are really 'unscientific', they remain a significant feature of PMS and one that sociologists need to understand both in their socio-logic and social significance.

Science and PMS – From Rationalization to Rhetoric

There are a number of points to draw out from this discussion. First, it is apparent from Weber's own account that there are more possibilities for PMS than disenchantment alone. Second, his focus on disenchantment seems to contravene his own prescriptions for the scientist in treating an ideal-type as an empirical description and in drawing moral conclusions from this account of the 'facts'. Third, his comparative studies of world religions provide the basis on which to undertake a re-crafting of rationalization. The first two points provide grounds to argue that his stress on disenchantment was made with deliberate rhetorical intent, which also helps to account for the otherwise somewhat puzzling fact that he is an advocate for disenchantment as a moral imperative – the appropriate way for scientists to conduct themselves – even though he appears to hold its prospect in pessimistic regard.

As seen, in addition to the standard view of disenchantment, Weber distinguishes at least two other possible prospects stemming from the Romantic spirit he detects amongst German youth: one that would direct the process of intellectualist rationalization into the 'irrational' spheres of art and love; and one that would precipitate a return to the 'war of the gods' as conflicting impersonal forces. The latter is especially significant as it seems to contradict somewhat the broader characterization of disenchantment: how is it that even depersonalized 'gods' (that is, value commitments) can retain any hold over the modern individual, if that individual believes solely in the scientific explicability of the world entailing commitment to formal rationality (Locke 2001b)? Now, it is not so much that an argument to this effect might not be made, but that it shows Weber's stress on disenchantment reaches beyond an analytical model towards empirical description, at least in the strong inference he draws regarding the meaning of science. In effect, he presents this as an empirical description about PMS to bolster the case for why scientists must act in one way rather than another as a matter of moral duty. Strictly, in his own methodological terms, disenchantment is a deduction derived from the ideal-typification of formal rationality; thus, it is an analytical extrapolation, not an empirical description. Moreover, to be true to the vocation of the scientist as Weber presents it, neither the model nor any extrapolations drawn from it should be used to intervene in practical affairs. However, he is not only saying, 'this is how the world is', but also, 'this is how it *should be* for the scientist'.

There is a parallel here with the Mertonian model of the normative system of science. Merton (1968a) identified an ideal-typical set of norms that he argued constitutes the necessary value-system for the production of valid scientific knowledge. Thus, it is not so much that he is describing how individual scientists do behave, but stating that this is what must be true about science as a social system if valid knowledge is to be produced. As a consequence, there must be a tendency for individual scientists to conform to these normative requirements otherwise science would not produce valid knowledge – and in contexts where the norms have not been followed due to political and economic interference (substantive rationalities), valid knowledge has not been produced (Merton 1968b). However, Mulkay (1976) argued persuasively that there is no independent means of determining if scientists do actually follow the norms, because all we have to go on are their self-descriptions, such as research reports, which invariably present their actions in line with the normative model. But it does not follow that this is actually how they acted. Moreover, scientists' actions can always be described in contrasting ways, as is apparent both from appeals made to alternative norms and in scientific controversies where one scientist's *Zweck* is another's *Wert* and vice versa. In the light of this, Mulkay (1993: 71) argued that the supposed norms of science should be treated by sociologists not as actual descriptions of scientists' actions, but as rhetorical characterizations or 'standardized verbal formulations'.

So too, I suggest, disenchantment and the characterization of science and formal rationality on which it is based. Like Merton's norms, Weber treats formal rationality as in effect the necessary value system of modern science, from which he derives disenchantment as the implication for PMS. However, his own discussion makes apparent that alternative actions are possible both by scientists (the 'big children') and by others (such as German youth). Thus, the characterization of science as purified instrumentalism should be viewed in the same manner as Merton's norms, as a standardized verbal formulation or rhetorical account. This does not mean individual scientists may not act instrumentally or that disenchanted outlooks may not be adopted by ordinary people, but rather that we cannot treat any such descriptions of actions or outlooks as definitive. They are instead to be viewed as versions that do specific rhetorical work. In effect, this is what Weber does in his lecture: he presents an argument. The argument is that scientists should act in one particular way rather than another, based in part on a claim about PMS that by his own account is disputable. Further, in so arguing, he breaches his own methodological principles, but does so with specific rhetorical intent to make the case for one way of doing science rather than another, a way he asserts follows from the characterization of science as pure instrumentalism.

I have belaboured this a little because it seems to me the standard view of rationalization in contemporary sociology, following primarily the influence of the Frankfurt School,[2] has been to treat disenchantment less as a rhetorical characterization than as an empirical description of both science and the mental state of the ordinary person. Instrumentalism has been assumed to be the prevailing attitude of mind in modernity, although it is often claimed this is now being undermined (Crook, Pakulski and Waters 1992, Chaney 2002, Hume and McPhillips 2006). The simple point, however, is that instrumentalism has always only been *one* available stance to adopt in modernity and Weber's own thesis can be re-crafted to provide grounds to argue that the assumption it has prevailed is not only a one-sided characterization, but if corrected, leads to a significantly different interpretation of contemporary social change. To show this, I now turn to brief consideration of two such interpretations: Lyotard's 'condition of postmodernity', and Beck's 'risk society'. I discuss them separately before drawing together some general points in conclusion.

2 There is no space here to elaborate, but the take up of the standard view of rationalization in the tradition of critical theory is apparent in Habermas' (1984, 1987) theory of communicative action (Locke 1999a: 175-184; see also Tiryakian 1992).

Beyond Postmodernity and Risk – Back to Modernist Rhetoric

Lyotard's (1984) thesis of the postmodern condition is useful to consider, because an argument has already been made that it presents a view of modernity in keeping with Weber's broad thesis (Gane 2004; see also Locke 1999a: 188-193). However, whereas Gane looks for continuities that stem into the postmodern, I will emphasize the superfluity of such a description, on the grounds that what Lyotard seems to be describing as regards science and PMS is consistent with a re-crafting of rationalization *within* the modern period.[3] The thrust of my argument is that there are unresolved ambiguities in Lyotard's description of both science and PMS that in part stem from his adoption of a view derived, if largely implicitly, from the Weberian model and that these disappear under the re-crafted version.

It is far from immediately apparent that Lyotard's analysis derives from a Weberian view of science; if anything, it seems to owe more to French sociology in characterizing the 'grand metanarratives' of modern science in broadly Enlightenment terms. Nonetheless, his (1984: 18-27) distinction between pre-modern 'narrative culture' and modern scientific culture as built around different kinds of language-game carries a hint of Weber's sharp distinction between facts and values in that, whilst narrative culture constitutes a self-legitimizing moral order, science requires that statements be legitimized independently of any such ground. However, as modernity has advanced, it has become apparent that this game is paradoxical, because the demand for legitimation must apply not only to statements that purport to describe the world, but also to statements that validate such descriptions – and so on to infinite regress. In addition, the two major Enlightenment metanarrative justifications of science, the Comtean account of the projected unification of knowledge and the Marxian account of universal practical empowerment have collapsed. In the first case, because any such unifying narrative must ultimately include itself making it self-legitimizing and thereby contravening its own language-game; and in the second, because this entails a moral justification that links science with an idea of a just social order and so appeals to criteria excluded by science.

In addition, although this collapse seems to be sufficient to have provoked the postmodern absence of an universal metalanguage, Lyotard points to a number of other features of the development of science that define this condition further, amongst which strong echoes of the Weberian model can be detected. But other voices can also be heard, leaving an ambiguous impression overall. So, in one respect, science is said to encourage the formation of new language-games, because it follows a logic of argumentation from first principles (Lyotard 1984: 43). Thus, scientists are continuously inventing new arguments and coining new theoretical and

3 Gane (2004) argues that constructions of 'before' and 'after' are strictly inapplicable as Lyotard challenges such temporal linearity. Maybe so, but the critical point remains that his characterization of the modern condition treats its rhetoric as real and, if we cease to do this, we no longer require a notion of the 'post'-modern at all (Latour 1993).

conceptual terms, thereby presaging the multiplicity of the postmodern condition. Against this, however, science has been delimited by what Lyotard (1984: 45) calls the 'performativity' criterion, which can be equated to instrumental rationality as it involves a calculative logic of technical correctness concerned with optimization of technological performance oriented to control over reality (Locke 1999a: 190; see also Gane 2004). But even here, developments in science such as quantum mechanics and chaos theory are undermining performativity, by showing reality to be ultimately unpredictable. The upshot is a new basis of justification of science through 'paralogy', generating new language-games as an end in itself. Thus, in this respect, science is presented as the basis of a new social model marked by the creation of new language-games in an unending, free-wheeling multiplicity of voices – the postmodern condition.

Now, one might read this analysis as providing some support for Weber's forecast of a resurgent 'war of the gods'. However, there are other features of Lyotard's discussion both of science and its social significance that raise doubts about this as being a novel development and, moreover, he presents quite contrasting characterizations at different points in his argument. So, while in some places science is presented as a model of postmodern paralogy, in others it is said to discourage such variety because the more radical the move in the language-game, the less likely it is to be well received by the established scientific community (Lyotard 1984: 63). This alone might make us doubt that science is really much of a model of postmodernity, but if we add in the view that it is assessed by performativity, it begins to look even less polyphonic and creative. A similar tension is found in his description of the relationship between science and society. At times, again in rather Weberian mode, modernity is described as a society dominated by monological performativity. At other times, however, a rather different view is given, as when he (1984: 25) writes of the language-game of science being separated from the 'general agonistics' of the wider society in relations of 'mutual exteriority', to the extent that the latter is still marked by the narrative form characteristic of pre-scientific culture. As an example, he (1984: 27-28) states that scientists 'recount an epic of knowledge' and 'play by the rules of the narrative game' in popularizing their discoveries to the media, rather echoing the 'dumbed down' view of popular culture discussed in Chapter 1.

These confusions are further compounded by another description of modern culture, as marked by an apparent multiplicity and heterogeneity of language-games of the sort that, one assumes, characterizes the postmodern. Modern culture is, he (1984: 65) says, 'a monster formed by the interweaving of various networks of heteromorphous classes of utterances'. So, why is this not already taken as 'postmodern paralogy'? This would give a quite different view of the relation between science and society, such that it is society that provides the model of the so-called 'postmodern' condition and not science at all. Here, it would be science that is set apart from a more widespread heteromorphic agonistics and science that is marked by performativity, not the wider society. Science would play the more limited and restrictive language-game, whilst the wider society is and has always been characterized by the multiple argumentation of paralogy. In which case, why do we need the concept of 'postmodernity'?

The concept of 'postmodernity' arises, I suggest, not so much because we need a term to describe a new social condition, but rather because social theorists have worked with a mistaken image of modernity. Lyotard's text is an unintentional documentation of a struggle between two contrasting rhetorical characterizations of science and its public meaning that need to be disentangled. The first comes from the standard view of rationalization and represents science as epitomizing an instrumental mode of reasoning that dominates the general societal outlook. This rhetoric is apparent when Lyotard writes of performativity characterizing not only science but the whole of society. The second characterization comes from elsewhere, but is not out of keeping with Weber's overall thesis, *if* we recognize that disenchantment is only part of the story. In this image, science is a multiplicity and so too is society. Science is seen as something that encourages ever new language-games and moves within language-games – in a word, science is argumentation. But society, too, is marked by such heterogeneity, incorporating a multiplicity of versions of reality and the rhetorics that support them, all caught up in a general 'agonistics' or argument. This generality is not a new thing, not a novel condition; it has been ever present in modernity (Latour 1993). Thus, in this second image, modern society is *already marked* by the argumentative multiplicity and heterogeneity that Lyotard reserves for postmodernity. Therefore, we do not need this notion. All we have to do is accept that the standard view of rationalization systematically misrepresents science and its public meaning as a consequence of Weber's rhetorical concern over the ethic of conduct of the scientist. This way of characterizing the relationship between science and society is only *one* possibility derivable from his broad thesis, which also includes, not least, Lyotard's 'postmodern' envisioning of the 'war of the gods'.

Interestingly enough, a very similar conclusion is reached from consideration of Beck's (1992) ostensibly rather different thesis of 'risk society'. Like Lyotard's 'postmodernity', 'risk society' offers a revised understanding of modernity to make sense of perceived changes in recent times, but unlike Lyotard, Beck is explicit in his use of Weber and so the argument can be made rather more directly. Beck argues that modernization is an uncompleted project of which the process of industrialization was only the first stage and we are now entering the second stage that he calls 'reflexive modernization' (or 'risk society'). These stages are defined by two forms of 'scientization', primary and secondary. 'Primary scientization' was characteristic of industrialism and is defined explicitly in terms of Weber's rationalization thesis: 'Max Weber's concept of 'rationalization' no longer grasps this late modern reality, produced by successful rationalization. Along with the growing capacity of technical options [*Zweckrationalität*] grows the incalculability of their consequences.' (Beck 1992: 22) Thus, it is taken as given that the first stage of modernization resulted from 'successful rationalization' understood in instrumental terms leading to the technologization of production. This was directed at wealth-creation, a by-product of which was a variety of 'risks' such as pollutants like urban smog, but these remained relatively excluded from dominant social groups. Thus, although quite visible, they affected mainly the poor and socially marginal, and could be justified as necessary for the general improvement in overall social well-being. Significant during this stage is that scientific scepticism was confined to 'external' matters, the 'objects of research' (Beck 1992: 14). So although scientists may have questioned ideas about nature, they did so within limits defined by 'technocratic and naturalistic' (Beck 1992: 24) categories of thought that generally prevailed throughout society.

Now, however, a new stage of 'secondary scientization' has been reached, which is marked by the growth of the 'suspicion of fallibility' about science (Beck 1992: 14). In part, this is due to the creation of new, invisible types of risk with universal impact such as radiation, the effects of which even the wealthy cannot escape or ignore. As a result, scepticism now extends to the internal relations of science, to the professional community itself; in other words, where once the capacity of science to provide valid knowledge of the world was accepted, this is now doubted – or in Weber's terms, it is no longer the case that people so readily believe that the world is knowable in principle through science. This, then, has led to the politicization of science manifest in, amongst other things, calls for its democratization (which, as seen in Chapter 1, are associated with some social scientific critiques of the PUS movement). Beck (1992: 14) incorporates into this account a range of other social changes involving wider social institutions such as work, gender and sexual relations, pointing to the growing freedom from 'the certainties and modes of living of the industrial epoch' that, together with the doubts about science, constitute the new state of 'risk' in which we now live.

However, similar unresolved ambiguities regarding the nature of science and its public meaning can be found in this argument as are discernible in Lyotard. First, regarding science, Beck characterizes this as marked by both a monologic of technical reductionism and a polylogic of scepticism and argumentation. Thus, science is held to be the source of *both* the industrial stage of modernization *and* the basis of 'reflexive modernization' – just as it is also the source of environmental problems and the solution to them. Second, with respect to its public meaning, society is said to be dominated by science through, in the 'primary' stage, monological technical/naturalistic categories of thought and, in the 'secondary' stage, reflexive scepticism towards such categories. The unresolved ambiguities here are illustrated by Beck (1992: 22-23) treating scientific risks as both real and as socially constructed, as in this passage:

> By risks I mean above all radioactivity, which completely evades human perceptive abilities, but also toxins and pollutants in the air, the water and foodstuffs, together with the accompanying short- and long-term effects on plants, animals and people. They induce systematic and often irreversible harm, generally remain invisible, are based on causal interpretations, and thus initially only exist in terms of the (scientific or anti-scientific) knowledge about them. They can thus be changed, magnified, dramatized or minimized within

> knowledge, and to that extent they are particularly open to social definition and construction.

Here, 'risks' seem to have a dualistic status being, on the one hand, things that are both beyond the human capacity to detect (they '*completely evade* human perception') and able to do things independently of people (such as '*inducing* systematic and irreversible harm'), and yet, on the other hand, dependent for their existence on human understandings (they '*only exist* in terms of knowledge about them') and open to human capacity to influence ('they can be changed' and 'are open to social definition and construction').

The ambiguities here regarding the ontological status of the phenomena in question and the epistemological status of our knowledge of them – that is, are they actually real or socially constructed – reflect the broader confusion in Beck's account over the nature of science and its public meaning (Wynne 1996). Like Lyotard, this confusion arises from attempting to accommodate two different images, one drawn from the standard view of rationalization and one taken from social constructionism. In the first image, science appears as a monolith characterized by a singular logic of instrumentalism that dominates the mental outlook of both scientists and ordinary people, manifested in an over-riding concern with technical efficacy and a belief in the capacity of science to explain the world fully. In the second image, however, science appears as a polymorph characterized by a logic of sceptical argumentation that, whilst still dominant, presages a new form of society in which ordinary people are doubtful about both its technical capacity to manage problems (not least those it has itself created) and its ability to explain the world fully. Thus, science is recognized as socially constructed and as providing only one possible (or a range of possible) understandings and orientations towards the world that are open to influence from external interests – and so we return again to the 'war of the gods'.

Conclusion

Lyotard and Beck far from exhaust the rich variety of social theorizing about the nature of contemporary social change in modern societies. Nonetheless, their views are to some extent indicative of a wider tendency to accept more or less as given that modernity is a scientific society in the particular sense derived from the standard view of Weber's rationalization thesis – a disenchanted society dominated by an instrumental mental orientation. Given this, the puzzle these theorists are presented with and which their theories are designed to solve is how this supposedly monolithical and monological scientized world has produced what now appears to them as a heterogeneous, multiple, agonistic world, in which science itself seems increasingly subject to critical doubt both from within and without the academy. The strategy adopted to solve this puzzle – to postulate a general change in the social condition of modernity, whether towards postmodernity or risk, as a direct result of internal changes within science then producing a new type of mental orientation in society at large – is unsatisfactory because, in effect, it does not solve the puzzle at all but simply restates it in a hidden form. The puzzle then becomes, how is it that what was apparently a monolithical mode of science transmogrified into a multiple, sceptical and argumentative mode of science? Moreover, if it was the case that people's mental outlook was dominated by disenchantment, why should they give this new mode of scientific reasoning any credibility whatsoever? How are they even capable of registering it?

A different solution is required. In the solution I propose, modernity is a society that has *never* been dominated by instrumentalism and disenchantment. The standard view of rationalization treats a rhetorical characterization as a real description. It neglects that Weber's thesis is an argument made partly to suit an ideal-typical analysis of intellectualist rationalization, with this itself used to justify a stern moral warning to the scientists of his day regarding the possible consequences of their not acting like pure instrumentalists in the pursuit of knowledge – a return to the war of the gods. Two things stem from this: Weber did not hold that the war of the gods had actually stopped in modernity, even if he may have thought it likely to be eclipsed by bureaucratization; and we should not take his characterization of science and its public meaning at face value, but recognize it as a deliberately accentuated one-sided version for the rhetorical purpose at hand. It is only one possible outcome of the process of rationalization as he analyzed it. A re-crafting of rationalization then involves recovering the other possible outcomes, not to abandon the notion of disenchantment, but to treat it as only one possibility amongst others and thus as itself part of the ongoing public argument about what science means. Our contemporary situation is a continuation of this ongoing argument. There is no reason to assume any fundamental change in our social condition in this respect, or that other social changes have been precipitated by professional, institutionalized science. Rather, science is itself caught up in the 'general agonistics' *as it has always been*; and science is itself caught up in the reflexive understandings, the practical sociological reasoning of everyday life *as it has always been*. In Weber's terms, we might call this the 'war of the gods'. In the next chapter, I attempt to chart the major strategies available to the warring factions using the insights offered by his ideal-typical historical analysis. This involves re-crafting the ideal-type to recover what is already implicit, that the model of formal rationality is really only another form of substantive rationality. Having recognized this, the 'informal' ways in which substantive rationalities, and formal and substantive irrationalities may also be present within science can be discerned and their implications for PMS outlined.

Chapter 3
Re-crafting Rationalization: Why Science Does Not Mean Disenchantment

In the last chapter, Weber's thesis of intellectualist rationalization was overviewed to bring out the wider range of possibilities for PMS than is commonly considered. The standard view of rationalization emphasizes disenchantment – the loss of belief in the unknowability of nature – as the characteristic mental outlook of the ordinary person in modernity directly resulting from the singular logic of instrumentalism informing science. This view is found in much contemporary social theory including Lyotard's thesis of postmodernity and Beck's thesis of risk society. However, in these theories the standard view is mixed up with a quite contrasting view in which science itself is characterized by a multiplicitous logic of argumentation, while PMS is both heterogeneous and informed by a reflexively critical attitude towards science. In this chapter, I try to show that this mix can be understood without needing to postulate any fundamental change in social condition, through a revised understanding of rationalization that re-crafts Weber's thesis to rescue the wider range of possibilities he downplayed in his concern over the proper moral commitment of the scientist. Of particular interest is the possibility of a continuing enchanted attitude towards science both at a general level and in such specifics as the theories and actions of scientists and their associated products. To bring this out, I draw on work by the sociologist, Harriet Whitehead (1974, 1987) and the rhetorician of science, Thomas Lessl (1985, 1989). Before I get to this, however, some re-crafting needs to be done.

From the overview of intellectualist rationalization a number of additional possibilities to disenchantment were identified. Alluding to the spirit of Romanticism, Weber (1948b) referred to one in the form of a revised 'war of the gods' albeit involving impersonal forces rather than the personalized entities of traditional beliefs. In relation to the broader discussion of intellectualist rationalization, this can be taken to refer to various substantive rationalities and 'irrationalities' drawn from other value spheres that might then inform both science and its public meaning. There is, however, an additional possibility that emerges from the disjuncture identified in Weber's account of Confucianism. This relates to his view of formal rationalization as involving both 'theoretical mastery' through conceptual clarification and abstraction of intellectual understanding, and 'calculation of means' involving a purely instrumental treatment of the relation between means and ends. In his general comments on the meaning of 'rationalism',

Weber (1948c: 293) states that although 'different' these 'belong inseparately together'. However, in his comparison of Confucianism and Protestantism, the critical distinction he makes between them seems to reside in just this separation: although both are disenchanted at the theoretical level rejecting magical understandings of the world, only Protestantism actively works to eradicate such understandings as a result of its instrumentalist social psychology. Thus, the historical occurrence of a situation in which the two aspects were separated raises the possibility of an alternative dimension to formal rationalization within science with, as shall be seen, significant implications for PMS.

Rationality and Irrationality in Law and Science

These various possibilities can be brought out further from considering Weber's (1968: 656-657) comments about the rationality of legal proceedings. He begins with an ideal-typification of formal rationality in which

> in both substantive and procedural matters, only unambiguous general characteristics of the facts of the case are taken into account. This ... can ... be of two different kinds. It is possible that the legally relevant characteristics are of a tangible nature, i.e., they are perceptible as sense data. This ... represents the most rigorous type of legal formalism. The other type ... is found where the legally relevant characteristics of the facts are disclosed through the logical analysis of meaning and where accordingly definitely fixed legal concepts in the form of highly abstract rules are formulated and applied.

He then identifies three types of deviation from this pure form: formal irrationality 'when one applies in law making or law finding means which cannot be controlled by the intellect'; substantive rationality where 'the decision of legal problems is influenced by norms different from those obtained through logical generalization of abstract interpretations of meaning ... includ[ing] ethical imperatives, and other expedential rules, and political maxims'; and substantive irrationality when 'decision is influenced by concrete factors of the particular case as evaluated upon an ethical, emotional or political basis rather than by general norms.'

I want to argue that a parallel can be drawn with intellectualist rationalization in the specific case of science, although to do so involves some re-working. The description of formal rationality above fits broadly with the formally rationalized ideal-type of intellectualist rationalization at least in respect of 'theoretical mastery' through precise and abstract concepts. In addition, Weber's comparison of the development of science in India and China with the west makes clear that he considered empirical experimentation – and hence 'sense data' – a crucial feature. Meanwhile, the three deviant types vary either because decision-making includes means that 'cannot be controlled by the intellect', or because of the intervention of substantive interests influencing decisions in respect of general norms or concrete particulars. This fits remarkably well with the kinds of internal characteristics of the scientific community described by science and technology studies (STS) and provides a basis for making sense of PMS in conjunction with the alternative scenarios outlined above. To draw out the parallel with science fully, however, requires a conceptual shift to take us away from Weber's troublesome terminology of 'rationality' to think instead of discourses and rhetorics. As argued in Chapter 2, the term 'formal rationality' invokes an implicit appeal to a socially transcendent conception of reason that is no longer tenable, specifically because it is asymmetrical (on this, see Bloor 1976, Mulkay, Potter and Yearley 1983, Ashmore 1989, Collins and Yearley 1992, Potter 1996: 25-41). Rather, we should think of a formalized discourse that provides scientists with 'standardized verbal formulations' (Mulkay 1993: 71), a set of rhetorical resources with which to characterize their own and other's actions and beliefs. Thus, in respect of science, I treat what Weber calls 'formal rationality' as just such a standardized discourse involving a characteristic set of rhetorical formulations.

In Weber's terms, 'formal rationality' in science involves 'theoretical mastery' and 'calculation of means'. In the analysis of scientists' discourse, meanwhile, reference has been made to a set of rhetorical techniques referred to as the 'empiricist repertoire' (Gilbert and Mulkay 1984, Ashmore 1989, Wooffitt 1992, Potter 1996, Locke 1994, 1999a, Kerr, Cunningham-Burley and Amos 1997). This includes three main things: the use of passivized linguistic formulations (basically, the 'passive voice'); the treatment of data as primary; and the treatment of laboratory work as standardized in form (Potter 1996: 153). There are some points of overlap here with Weber's formal characterization. For one, 'theoretical mastery' involves the development and application of abstract concepts which Weber associates with depersonalization. This fits with features of passivized grammatical forms such as nominalization (Hodge and Kress 1993) such that the passive voice is employed to signify conceptual abstraction (Halliday 1998; see also Billig 2008a, 2008b, Fairclough 2008a, 2008b). For another, the treatment of data as primary fits with the 'most rigorous' aspect of formal rationality in law involving reference to 'sense data', which in turn dovetails with Weber's stress on scientific experiment.

There are some differences, however, so I suggest that Weber's characterization can be viewed as referring to a broader set of rhetorical features that may be used in the formal representation of science of which the empiricist repertoire is a subset. The empiricist repertoire refers principally to features of scientists' discourse employed in such formal contexts as scientific papers, so it may be that in broader contexts of self-representation, especially those involving non-scientists, scientists might deploy a wider range of verbal formulations. Some evidence for this comes from the discussion in Chapter 1 regarding the representation of the non-scientific 'other' as 'irrational'. Concomitantly science is often presented as epitomizing rational thinking which in turn may be explicitly associated with the calculative reasoning of mathematics (for example, Dunbar 1995) and used to distinguish science from ordinary 'commonsense' (for example, Wolpert 1992; see also

Derksen 1997, Lamont 2010). Further, the contrasting possibilities provided by both rationalist and empiricist features provides scientists with a highly flexible set of resources that may be mobilized to construct quite contrasting boundary lines with different social groups on different occasions, as in the contrast between 'pure' and 'applied' science (Gieryn 1983). Thus, 'theoretical mastery' and 'calculation of means' can be viewed as features of the formal discourse of modern science that provide scientists with a diverse and flexible set of rhetorical resources of self-presentation, regardless of whether they actually describe qualities of individual scientists in terms of their actions and beliefs or mental orientation.

A further indication that calculation of means provides such a resource comes from considering how Weber's sense of 'formal irrationality' might apply to science. In the case of law, this refers to means of decision-making not controllable by the intellect. Weber (1968: 656) specifically mentions 'oracles', something we might not expect to find in modern science, but if we think in terms of rhetorics of argumentation in the context of scientific controversies then clearer possibilities for striking a parallel arise. A key feature of such controversies is that they shift very quickly to a methodological level, as shown in Collins's (1985, 2004, Collins and Pinch 1993) discussion of the controversy over gravity waves. A similar move is found in cases involving 'fringe' science such as cold fusion (Collins and Pinch 1993, Gross 1995, Taylor 1996), homeopathy (Picart 1994, Brossard 2009) and parapsychology (Collins and Pinch 1979, 1982, Collins 1985; see also Hess 1993), in which questioning methodology is used as a means of boundary demarcation. Parapsychology is especially pertinent as the issue of the extent to which the phenomena in question are no more than chance occurrences is a central consideration, although there is a general sense in which any supposedly uncontrolled aspect of method is almost by definition arbitrary. It is not an exaggeration to say that for 'debunkers' the methods of parapsychologists are comparable to those of oracular divination – one might just as well read the entrails as try to read minds through Zener cards. Thus, absence of sufficiently calculatively controlled means is treated as 'irrational' in terms of the 'formal' logics of science. Here again, Gilbert and Mulkay's (1984) account of interpetative repertoires is relevant, specifically what they call the 'contingent repertoire', which consists of a range of resources employed by scientists to undermine the legitimacy of opposing viewpoints including various types of 'cognitive error' (Locke 1999a; see also Mulkay and Gilbert 1982). This includes (purported) errors of reasoning as well as methodological matters, but broadly constitute what might be considered 'formal irrationality' within science. Thus, 'formal irrationality' refers to one set of rhetorical resources that scientists can deploy in order to undermine the credibility of opposing viewpoints on cognitive and/or methodological grounds, thereby constructing boundary demarcations from 'pseudo-scientists' or 'unorthodox' (Lessl 1988) groups and individuals by questioning their reasoning or research methods.

Such resources then are clearly of significance in PMS as also are substantive rationalities and irrationalities. In law, for Weber, these refer to 'non-logical' features used to make decisions and involve respectively general norms or concrete particulars. Thinking of this in discursive terms again allows parallels to be drawn with the rhetoric of scientists in the context of controversies. Both refer to the presence of interests other than those appropriate to the formal procedures of the sphere of action intervening in decisions, interests that may be of either a social or personal nature. The parallel with other features of the contingent repertoire is marked. In addition to 'cognitive errors', scientists in controversy commonly refer to a variety of intervening, ostensibly 'non-scientific' interests as a means of accounting for opponents' 'errors'. These come in two broad types: those of a group and of a personal nature (Locke 1999a: 123). The former refer to types of social interests, such as economic, political or ideological factors supposedly interfering with the purity of a scientist's work. This might include for example accusations that research agendas are shaped by demands from industry, government or religion, and that theoretical frameworks reflect forms of group commitment such as those of class, gender or ethnicity. In Weber's terms these are substantively rational matters involving general norms drawn from the ethics of conduct of other value spheres. Personal contingencies, meanwhile, include a variety of more individualistic interests and motivations taken to account for a scientist's errors in a specific case, such as personal bias, fraudulent behaviour and so on. Broadly then, this has some equivalence to Weber's category of substantive irrationality. The degree of fit in either case may not be entirely perfect, but there is nonetheless a sufficiently striking overlap to merit some further reflection especially regarding social interests/ substantive rationality.

The focus on social interests in science was a principal concern of the Edinburgh School following the Strong Programme in SSK (Bloor 1976, Barnes 1977, Barnes and Shapin 1979, Barnes and Edge 1982, Barnes, Bloor and Henry 1996). I want to suggest that it is such interests that Weber had in mind when he referred to the 'war of the gods' in relation to modern science and this raises the question of how we are to think about such substantive rationalities sociologically. Specifically, are they a resource or a topic (Zimmerman and Pollner 1971)? Gilbert and Mulkay (1984; see also Mulkay 1981, Mulkay, Potter and Yearley 1983, Potter 1996: 17-41) argue that sociologists should not treat social interests as 'causes' of scientists' actions and beliefs, because scientists themselves already do as much in their accounts of error. They should not then be used as a resource of sociological explanation, but treated as a topic to be explained because they are part of the phenomenon we are attempting to describe and understand. If we use them as an explanatory resource, we are effectively siding with one side or other in what itself is a matter of dispute amongst the participants involved. If, for example, we were to claim that proponents of 'intelligent design' (ID) as a scientific theory are actually expressing a religious belief and so acting in religious interests, then we are accepting as valid the views of their critics who themselves assert as much (Fuller 2007), when what we should be doing is studying this claim as a feature of the dispute regardless of whether or not we think it valid (Locke 1999a). The attribution of social interests is part of the socially available means of disputation and to be studied as such.

So, it is not that we might necessarily be wrong to say that belief in ID reflects religious interests, but that proponents of this belief dispute this is so is part of the phenomenon to be understood. For sociologists to assert they are wrong would not only intervene in the argument in favour of one side, but also imply some form of conscious or unconscious deception on the part of ID proponents and thus to advance a type of ideological analysis of the sort rejected in Chapter 1 on the grounds that it treats people as cultural dopes. People may deceive and they may be deceived, but accusations of deception are also a means of undermining credibility and as such suasive devices. In a situation where comparable rhetoric is employed by both (or all) sides, we are in no position to decide who is right or wrong, since all we have to go on are the accounts given by the participants. Thus, it is in keeping with Strong Programme tenets of 'symmetry' and 'reflexivity' (Bloor 1976) to treat social interests/ substantive rationality as a topic and not a resource of explanation. To do otherwise is to engage in victim sociology treating people as subject to hidden social forces conspiring against them.

In my view this applies to PMS as much as to internal scientific controversies and this is where theorists such as Lyotard and Beck are unhelpful, because they both treat critique of science as a resource of explanation to account for the supposedly new social condition they describe. But because they also see science as the defining feature of the preceding condition of modernity, they are unable to account for where critique of science itself has come from – except from within science. Hence, they present a muddled view of both science and its public meaning: on the one hand, science is marked by a singular logic of instrumentalism and PMS by a condition of disenchantment; on the other hand, science is marked by a multiplicitous logic of argumentation and PMS by heterogeneous critique of science. They treat the latter as a new condition but also use it as the means to explain how this condition has come about.

To straighten out the muddle we have to treat critique of science simply as a topic of study. It does not explain social change simply because it is not a new phenomenon, neither as a feature of the internal workings of science nor in terms of its external public meaning. Here the analysis of scientists' discourse is a powerful corrective because it shows that scientists have always worked with contrasting rhetorics including those that, with Weber, we can call 'formal' and those we can call 'substantive'. The former makes reference to a range of empiricist features that accentuate the objectively factual grounding of scientific knowledge and a range of 'logical' features that accentuate its theoretical and methodological rationality; the latter refers to a diverse range of interests of both a general normative (social) and more particularized (personal) nature – the 'war of the gods'. Such rhetorics are present within science and *must always have been,* because they provide the resources of internal scientific argumentation that uphold the assumption of the existence of a singular world in the face of contending interpretations of that world and, just as important, they provide the resources of external boundary demarcation whereby the scientific community can uphold its distinctiveness from alternative and competing worldviews or other versions of the 'life of the spirit'. But it does not follow from the public self-representation of science by scientists as an uniform and unified monolith – a formally rationalized, disenchanted instrumentalism – that this is how it has always and everywhere been understood by other social groups and individuals. This is partly because scientists themselves present different versions of science in different times and places, *albeit any given version is routinely presented as the single, definitive version of science* and partly because people have available a range of other resources drawn from other value spheres and alternative worldviews. The mistake made by many social theorists is to treat one version of science – the disenchanted, instrumental version – as definitive and then view others as 'deviant', erroneous, or in the case of Lyotard and Beck, as 'new'.

Support for a view of science as multiplicitous also comes from the study of the rhetoric of science, a field with many similar concerns to SSK that developed largely independently in the United States at around the same time (Nelson, Megill and McCloskey 1987, Fuller 1993, Taylor 1996, Harris 1997). Thus, rhetorians of science have stressed the emergence and stabilization by the early twentieth century of a characteristic generic form used in the formal presentation of (natural) scientific research papers that broadly can be characterized as empiricist/'formally rational' (Bazerman 1981, 1987, 1988, 1997, Dear 1991, Nyhart 1991; see also Ziman 1968, 2000). However, they have also studied wider contexts of scientific argument that in general emphasize both the multiplicitous character of scientists' rhetoric and public responses to it (Campbell 1975, Bytwerk 1979, Fahnestock 1986, 1997, Gross 1994a, 1994b, 1996, Lyne and Howe 1997, Waddell 1994, 1997, Reeves 1997). These point to the existence of a variety of already existing contending substantive interests and forms of critique of science arising from diverse modern and pre-modern sources. Amongst these would be included those mentioned by Weber, notably Romanticism in the 'aesthetic' value sphere (hence the 'two cultures' – Snow 1964) and the revised 'polytheism' informing popular culture that has flourished into an immensely rich and diverse 'cultic milieu' (Campbell 1972) or 'occulture' (Partridge 2004), as well as others such as science fiction and even traditional religion – notably forms of fundamentalism such as creationism (Lessl 1988, Prelli 1989a, 1989b, Taylor 1996). Examples of these are discussed further in Parts II and III, so suffice it to say for now that they each provide rhetorical resources of their own including amongst other things: alternative ethos (representation of self), pathos (representation of the audience) and logos (knowledge claims) (Edmondson 1984); forms of figuration (preferred metaphors, metonyms and synecdoches) (Lessl 1985, 1989); and enthymemes (Locke 2002; see also Gross 1996).

This does suggest that widening familiarity with scientists' rhetoric has provided opportunities to articulate these and other forms of critique in the terms of science itself, employing similar rhetorical techniques and ostensibly using scientific expertize against itself (Nelkin 1992a). Moreover, this activity has a reflexively self-constituting reinforcement: the more critique of science in the terms of science appears, the more becomes possible because more people learn the 'rules of the game' and the more the presence of internal expert disagreement becomes apparent. In addition, the 'core set' (Collins 1985) of those caught up in scientific controversies involves a wider range of actors to the point of becoming 'potentially infinitely extendible' (Michael and Birke 1994: 83-84) and enabling opportunities for alternative rhetorics to be mobilized (for example, Waddell 1997). But in basis there is nothing new about this; it simply continues the conversation with and about science that modern culture has always had, one outcome of which is a rich and widening stream of 'hybrid' mixtures that meld science with various ('non-scientific') 'others' in an ongoing inventive bricolage. From the point of view of 'formal' scientific discourse this might seem to be (or at least is represented as) an increase in social critique of science – a growth of 'anti-science' (Holton 1992, 1993). But for sociologists to view it this way is asymmetrical, accepting one 'formal' version of social reality as purely descriptive rather than itself part of the phenomenon to be sociologically understood. This brings us back to the arguments considered in Chapter 1. The view built into the deficit model of PUS, which is essentially educational (albeit supplemented by the 'entertainment' mode of PEST) is that the more science people know ('understand') the more they will love it (Turney 1998b). However if teaching people science also means teaching them, if only implicitly, the rhetoric of science, then the more science they learn the more informed in the arts of scientific argument they become. Thus, contrary to the standard view of rationalization as a 'one-dimensional' instrumentalism (Marcuse 1964), a rhetorical approach suggests otherwise because however monolithically represented, the rhetoric of science necessarily incorporates its other(s) (Billig et al. 1988, Billig 1996; see also Mellor 2003). Thus, even a disenchanted rhetoric of science provides resources for the articulation of alternatives to disenchantment, as is further examined in Part III. But in any case there is good reason to dispute that science has ever presented a monolithical disenchanted public image and this brings me to the disjuncture in Weber's view of formal rationalization.

Enchanted Science

As seen in Chapter 2, this disjuncture concerns the possibility of separating the two aspects of formal rationalization, 'theoretical mastery' and 'calculation of means'. In relation to the modern west, Weber asserted that the two belong 'inseparately together' and constitute the unique character of its intellectualist 'spirit', both disenchanted and instrumentalist. However, as regards science, a case for claiming otherwise can be derived from Whitehead (1974, 1987). It needs to be stressed that Whitehead's discussion of Weber employs somewhat different terms to those I have been using and so there is a measure of reading into her discussion in what follows. Nonetheless, the critical point of her argument that is of most importance concerns essentially the same disjuncture. She (1974: 552-553) refers to this as a 'critical ambiguity' in how Weber saw rationalization affecting religion: on one side, the 'magical'/charismatic is viewed as a quality of 'extraordinary power' attributed to a person or object; but on the other, it is a spiritual realm abstracted from the charismatically endowed and taken to exist in its own right as 'a dimension of experience'. In her view, accounts of the rationalization of religion in contemporary society tend to conflate these two aspects such that disenchantment is associated with 'not just the elimination of magical attributes ... but the restriction on the accepted modes of experiencing that accompanies this'. Hence disenchantment is linked to secularization in the specific sense of an orientation of mind that confines experience to 'this world'.

In the light of the foregoing discussion, I take Whitehead to be referring here to the distinction between 'theoretical mastery' and 'calculation of means'. As seen in Chapter 2, in Weber's discussion of Confucianism 'theoretical mastery' is associated with the 'elimination of magical attributes' (that is, disenchantment), but this did not restrict accepted modes of experience at least outside the social context of the mandarin; thus, the 'ecstatic' teachings of Taoism were tolerated. In the west, however, he argued that Protestantism did restrict the accepted mode of experience to that associated with an instrumental, calculative orientation and it sought to transform the world accordingly by eliminating magic as an abstracted realm of potential experience as well as an attribution of individuals and objects (priests and their symbolic trappings, etc.). So, in making clear the ambiguity, Whitehead is then able to argue that the rationalization of charisma involves more than just a singular process of disenchantment, but also 'a gradient of abstractions upon, or objectifications of, extraordinary power'. From this, she (1974: 553) then says:

> The process of rationalization, as it affects religion, is not simply one of the replacement of the magical (extraordinary power) with the scientific (ordinary power) ... It is, on a more fundamental level, the transformation wrought upon concepts of the charismatic as man [*sic*] seeks to fashion from them an over-arching and universal system of meaning.

This means that the 'extraordinary' is not necessarily displaced by disenchantment but rather progressively 'translated into the remote distance' (Whitehead 1974: 554), which leads to a highly significant re-visioning of science and its social meaning. Rather than seeing science as necessarily disenchanting, it can itself be seen to be informed by the charismatic. This is especially apparent in positivism, which

> bears witness to the fact that science served as a medium for the rechanneling of religious energies, for in it we recognize both an attitude of adulation toward technologic possibilities and an attempt at a comprehensive understanding of the human situation rather than the only partial understanding that science is legitimately able to provide. (Whitehead 1974: 555)

In a word – although Whitehead herself did not use it – science is *enchanted*.

Further, the view of positivism this gives makes an important contrast to Lyotard. The manifestations of enchanted science Whitehead describes are the same as the 'grand metanarratives' Lyotard distinguishes: practical empowerment (at least partly through technology), and universal knowledge. However, because Lyotard accepts the restricted notion of rationalization that associates it with instrumentalism ('performativity'), he sees the enchanted character of these narratives as working only to undermine the legitimacy of science. Consequently, he interprets their continuing abstraction as a collapse, leading to an entirely contrastive condition in which a monolith is replaced with a polymorph – heterogeneous postmodern 'paralogy'. Whitehead's revised understanding of rationalization presents a different possibility. Pointing to contemporary interest in mysticism and the occult as 'a fresh exploration of the charismatic realm', she (1974: 556) argued that a 'sophisticated' Weberian view would see such developments as continuing the rationalization process to a further level of abstraction 'away from the attachment to concrete symbolism' associated with institutional religion towards an interest in the nature of symbolization itself. Hence, in various 'New Age' movements there is an interest in identifying what are taken to be underlying similarities of reference and meaning behind the symbols and teachings of all beliefs, including science (see also Main 1999, Besecke 2001). In Whitehead's (1974: 556-557) view such developments mark a 'prophetic breakthrough' signalling 'a cultural advance in the direction of greater rationalism ... ushered in by an upsurge of charismatic calling-to-order.' Nonetheless, like Weber in his view of German youth, she also expressed concern that the 'prophetic energies may become dissipated in a pursuit of the "gods of the moment",' that is, 'all those concretely symbolic, imminently distracting and divisionary dinosaurs of old-time religion and the Occult.'

Whitehead applied this argument to science fiction and Scientology both of which I return to in Part II, where I also look at a form of 'old-time religion', creationism. My intention is to further develop her alternative take on rationalization in relation to the foregoing discussion of the rhetoric of science. In particular, the view she provides of enchanted science *within* the logic of rationalization can be given sharper focus through linking it to Lessl's (1989) analysis of the 'priestly voice', as a critical feature of the rhetoric employed in science popularization (see also Campbell 2001, Mellor 2003). This refers to a mode of speaking that, at least by analogy, links traditional religion and modern science. Lessl (1989: 184-185) contrasts it to the 'bardic voice' that speaks in the 'profane' tongue of the ordinary people. The priestly voice is 'sacred' taking a 'didactic' form and coming from an elite group who claim that it originates from some extraordinary source 'as revelations of spirit or nature'. As such, it 'descend[s] from above as an epiphanic Word, filled with mystery and empowered with extra-human authority'. Thus, it refers to a level and form of reality that most members of society cannot hope to attain, but which is nonetheless presented to them as foundational. As such, it is 'at once both near and remote' offering a 'sense of identity with respect to the wholly other, the gods or the cosmos at large'. So, it defines the cosmological order and situates humanity within this.

However, this sets up a tension that priestly discourse needs to manage and for which it employs a characteristic rhetorical form, the key features of which are discernible in popularized science as much as traditional religion. From the perspective of such elite groups, the problem is how to maintain control over their specialized discourse given that it performs this wider social function. Their specialized discourse belongs solely to them (see also Sharrock 1974), but it has to be made publicly available to ensure their definition of reality prevails and the legitimacy of their authority is maintained. In addition, there is the need to recruit new disciples from the 'profane' social world and for this priestly discourse has to be translated into the vernacular. But this means it may come to be considered a public resource available to all. As Lessl (1989: 184-185) put it, the priestly voice serves 'a missionary purpose', which on the one hand must make 'its esoteric concepts meaningful without overreaching the linguistic limits of an initiate audience', but on the other hand must also 'maintain the sacredness of certain institutionally owned elements – symbols, rituals, and incantations closely identified with the ethos and, especially, the gnosis of their culture'. These are 'inherently conflictive obligations': how do you translate the esoterica of technical scientific discourse into everyday language without losing control over it (see also Broks 2006)?

As already suggested, the short answer is you do not. Intellectual entrepreneurs are always a threat and so there is a continuous need for boundary-work (Derksen 1997), including various forms of 'debunking' for which science has its formal rhetorical techniques derived from 'theoretical mastery' and 'calculation of means' – reason and instrumental empiricism. These provide one set of rhetorical techniques to police the boundaries, but Lessl (1989: 188) identifies a further feature involving synecdoche:

> Priestly rhetoric is synecdochic to the extent that in it we find an institution portraying its particular ethos as the very essence of humanity. All the paths of history in the priestly reading ... lead ... to their historic culmination in contemporary science. The priest interprets all experience in terms of the specialized priestly subculture. The part stands for the whole.

Thus, not only is there ongoing reconstruction of the history of science to show that it necessarily leads to whatever current truth is held (see also Kuhn 1970), but the whole of history is also narrated as a story of science – as can be seen for example in evolutionary psychology (for example, Dunbar 1995; for a precursor, see Popper 1972, and see also Fuller 2007). This is also a promise to people 'of *what they might become*' (Lessl 1989: 188) not merely in the sense of converting to the scientific cause, but also in the sense of a projected future for humanity as a whole – and one might think here of the promises held out for the Human Genome Project (Gross 1994b, Nelkin and Lindee 1995, van Dijck 1998), or the various forms of 'cyberbole' trumpeted in recent times (Woolgar 2002a; see also Davis 1998, Escobar 2000).

Thus, using synecdoche the priestly-scientist provides an essentializing definition of the nature of humanity in relation to the totality of the scientifically constituted cosmic order. This then provides what Lessl (1989: 190) calls 'consubstantiality between scientist and nonscientist', presenting both as different aspects of the same basic 'substance' or single being – so although we are different, we are really the same. Equally, whilst science in general stands for humanness in particular, this priestly discourse also makes 'humanness in general to stand for science in particular', through 'draw[ing] out those aspects of the ordinary mind that already evince a scientific character. Those scientific *parts* of all humans – rationality, inquisitiveness, skepticism – are made … to be the defining features of the *whole* species.'

Overall, then, the synecdochic connections work in two directions: '(1) Elements of the scientific that overlap with ordinary experience are drawn upon in an effort to humanize the otherwise alien world of scientific objects, and; (2) elements of ordinary human experience that coincide with specialized scientific norms and values are reconstructed so that they achieve scientific meaning.' (Lessl 1989: 190-191) An illustration comes from Carl Sagan's early 1980s television series, *Cosmos*, where he (quoted, Lessl 1985: 181) used the now widely popularized image of 'star stuff' (Gribbin 2000):

> The surface of the earth is the shore of the cosmic ocean. From it we have learned most of what we know. Recently, we have waded a little way out, maybe ankle deep, and the water seems inviting. Some part of our being knows this is where we came from. We long to return, and we can, because the cosmos is also within us. We are made of star stuff. We are a way for the cosmos to know itself.

The imagery of 'star stuff' defines a point of identification with everyday discourse, humanizing science even as it scientizes humanity. In one direction, the sacred cosmology of science, with its technical account of 'stars' as the source of the atomic building blocks of complex matter through nuclear fusion, is humanized by connection to the material substance of human bodies, themselves taken to be the substrate of consciousness and thus who and what we essentially are; and in the other direction, the profane discourse of the commonsense world is employed through ordinary knowledge of 'stars'.

There is, however, another aspect to this that Lessl does not mention, given that stars in ordinary discourse can signify 'magic', a dimension of enchanted experience. Thus, describing us as 'star stuff', draws on this enchanted sensibility investing it with purposeful meaning and direction defined through scientific cosmology. This needs to be recognized as it opens up two contrasting interpretative possibilities. In one, the prospective narrative of humanity coming to know the stars scientifically and thereby completing the cycle of a cosmic journey can be interpreted as disenchantment – it removes the magic because the stars themselves, treated metonymically as the whole cosmos, become known and demystified. But in a contrary interpretation, it is the mysterious nature of the stars that invests this prospective journey with a sense of awe and cosmic wonderment. Thus, while in the one case the activity of scientizing humanity may disenchant, equally in the other the activity of humanizing science may enchant. As Lessl tells us, the transference proceeds in *both* directions; it is this that is overlooked in the standard view of rationalization.

Here then a link can be made with the earlier discussion. In Whitehead's revised account of rationalization, science itself is invested with a charismatic quality, one that in Enlightenment discourse promised both complete practical control (calculation of means) and comprehensive universalistic knowledge (theoretical mastery). Meanwhile, the rationalization of religion has led to a search for higher level comprehension of symbolic meanings in an understanding of the nature of symbolization itself. This has given rise to attempts to articulate a new language of the transcendent, one that seeks to go beyond the division between science and religion (see Chapter 5). However, this is partly made possible because science itself presents enchanted outlooks through the workings of the priestly voice. The priestly voice articulates the charismatic within science by presenting it as a total cosmological vision, a definition of the whole of reality within which humanity is defined and located. This enables *both* a discourse of instrumentalized disenchantment *and* a discourse of enchantment, with a higher level morality that proposes to transcend the 'war of the gods' through its unifying vision of humanity within and as the cosmic whole. Thus, as Lessl (1989: 191) points out, the collective pronoun 'we' is an important feature in constructing identity between science and the public, with which the scientist-priest claims to speak on behalf of all humanity, even though scientists are only a minority and only a minority of them are directly involved in space science, or genetics, or cybernetics, or whatever other part is being presented as defining the human whole on any given occasion. However, because science is presented as collectively owned and because it is invested with a quality of enchantment which may be accentuated in the transference of its transcendent vision(s) to the wider public (see also Fahnestock 1986), the opportunity – and the rhetoric – is then available for members of the public to use it as a resource in the construction of alternative forms of transcendence.

There is then a paradox of consequences to the rationalization of and by science: the more science becomes established as the legitimized discourse of reality definition, the more it facilitates the construction of alternatives. Further, such alternatives may move in two contrasting directions either towards disenchantment or towards enchantment. In either case, science provides rhetorical resources that can be turned to assist the activity. Or, to put this around the other way: disenchantment and enchantment define a central and constitutive dilemma of science in respect of its public meaning. On one side, science is disenchanted and disenchants; on the other, it is enchanted and enchants. The two together more fully define PMS in modernity and so provide two seeds of arguing and thinking that have flowered, at least since the Enlightenment, into an elaborate and diverse array of 'standardized verbal formulations'. These are both instrumental and value-laden, empiricist and contingent, calculative and theoretical, monolithic and polymorphous, technicistic/ naturalistic and non-technicistic/ non-naturalistic, and a yet wider heteroglossia of images and figurations, enthymemes and synecdoches. There are also elaborate and diverse visualizations, both 'black-boxed' (Greenberg 2004) and polysemic (Hüppauf and Weingart 2008; see also Locke 2009a), and multiple images of the scientist from 'mad' to 'saviour' with numerous variations and combinations between (Campbell 1975, Basalla 1976, Lessl 1989, Tudor 1989, Toumey 1992, Haynes 1994, 2003, Jancovich 1996, Jones 1997, Skal 1998, Turney 1998a, Jörg 2003, Weingart, Muhl and Pansegrau 2003, Locke 2005). These articulate and partly define the pervasive ambivalence to science in modern culture (Wynne 1995; see also LaFollette 1990), an ambivalence that is far from new. It is rather embedded in scientific discourse as an outcome of its rationalization that presents a purified, value-free, universal comprehension that is at once the promise of human transcendence and the threat of dehumanization.

This, then, is a re-crafted rationalization. Where Weber stressed disenchantment, regretting the eclipse of charisma and the tragedy of instrumentalism but fearing more a return to the 'war of the gods', I think we need to see that in this respect his worries got the better of his analytical insight. But the insight afforded by his analytical framework can be recovered and taken forward. I have tried to show that his model can be used to help make sense of both the internal workings of scientific controversy as well as provide a basis for a fuller understanding and further study of PMS. This is what I take forward in the rest of the book. In Part II, I focus on the dilemma of disenchantment versus enchantment, building out from Whitehead's view of the continuing rationalization of the charismatic as a process of increasing abstraction and Lessl's analysis of the synecdochic rhetoric of the scientist-priest. I apply these ideas to examples of both 'old-time' and 'New Age' religion – creationism/ ID and Scientology respectively – to explore how rationalization as a process containing both disenchanted and enchanted science works through. Specifically, I focus on how rhetorics of both disenchantment and enchantment – the 'formal' rationality of science – are employed to ostensibly 'non-scientific' (or 'substantive') ends, to construct religious cosmologies that involve the invention of an identity that is both scientized-humanity and humanized-science and which I call an *enchanted self*.

Continuing this theme, I then examine how disenchanted and enchanted rhetorics are played out in the case of science fiction, something mentioned by both Whitehead and Lessl as a site where charismatic, priestly discourse is articulated. Although neither specifically mention superhero comics this is my focus as I want to show that, even in this most culturally downgraded of contexts, the constitutive dilemma of science is manifest – and arguably, it is this lowly status that provides the best evidence against a monolithical understanding of science as disenchanting. After all, one might expect the supposedly meanest of cultural contexts to exhibit the barest face of hegemonic power; certainly this would be the conclusion of an old 'mass culture' perspective (Adorno 1991). But I will argue the very opposite, that superhero comics are a site of sophisticated cultural struggle over the meaning of science and one manifestation of the modernist urge to articulate a charismatic transcendence that incorporates but goes beyond science. Thus, the constitutive dilemma of science informs the ordinary practical sociological reasoning at work in superhero comics, as the rationalization of charisma is worked out in the construction of a coherent 'reality' in which science and magic co-exist. This is done through a specific usage of the documentary method of interpretation (Garfinkel 1967) known as 'continuity'. The ethnomethodological note struck here is then picked up in Part III where the focus shifts to disenchanted rhetoric, but with a view to arguing against instrumentalism as its single mode. Instead, I consider how a disenchanted orientation works through into the logics of mundane reasoning (Pollner 1987) leading to a further mode of rationalization that goes beyond anything so far considered – a paradox of consequences that I call mundane mysteries. But more on that after the enchanted self.

PART II
Enchanted Science

Chapter 4
'Fearfully and Wonderfully Made' – Designing an Enchanted Self

In Part I, an argument has been set out to the effect that a re-crafted version of rationalization points to the possibility of enchanted forms of science being a central feature of PMS. Now, in Part II, I will begin to bolster the case for this view by looking at some forms of such enchanted science in contemporary society beginning with religion in this chapter and the next, and concluding with science fiction and fantasy in Chapter 6. Religion is one of the major social contexts in which enchanted forms of science are to be found. In one way this is unsurprising since, as seen in Chapter 2, religion is the principal site in which the charismatic has traditionally centred; indeed from a Weberian point of view what we call 'religion' is substantially the unfolding of the charismatic as a discursive order. However, given that in the standard view of rationalization religion is supposed to be displaced by science intellectually through disenchantment and functionally through the differentiation of its traditional ministries such as education and counselling, that contemporary religious discourses draw upon science might seem very surprising. Of course, it is also to be expected that such secularizing processes stimulate reactions from established faiths in the form of 'resistance or accommodation' (Dawson 1998a, 1998b), whilst various new beliefs, albeit fragmentary and transitory are concocted to fill the gaps left by the decline of established religious authority (Abercrombie et al. 1970, Wilson 1976). However, these standard sociological terms are insufficient to account for religious groups advancing not merely accommodations of science, but outright challenges to its established position and doing so not so much in the way of resistance, but on what are presented as 'scientific' grounds. This is puzzling precisely because science is supposed to disenchant, to erode the ground from under the charismatic and replace its sense of mystery and magic with firmly established this-worldly instrumentalism. If this is what science does, how can it be claimed that it actually shows the truth of religion and even the validity of charismatic wonderment?

From the standard view of rationalization, as much as from the point of view of many in the established scientific community, such arguments are untenable and patently false. As such they have been treated sociologically as examples of the 'sociology of error' (Bloor 1976) to be accounted for by such social factors as the consequences of structural 'marginality' (for example, Nelkin 1982) and the breakdown of traditional community (for example, Wallis 1976). These are examples of victim sociology, a form of sociological reasoning in which ordinary members of society are seen as subject to large-scale social forces beyond their control that undermine the 'ontological security' (Giddens 1991) provided by established ways of life, resulting in pathological social conditions and mental states such as 'anomie', 'alienation' and 'anxiety' (or, more fashionably, 'risk'). The problem with such accounts is two-fold: first, much like the theories of mass culture encountered in Chapter 1, they effectively treat people as cultural dopes – passive, incomprehending dummies unable to act effectively for themselves; and second, as argued in the last chapter, their notion of rationalization is based on only one version of the discourse of modern science that overlooks the presence of enchanted versions in which science itself becomes a focus for the development of charismatically-imbued visions. The weakness of the first position is that it ignores the capacities people have to develop their own critical understandings of their social world and to act on this basis; and the weakness of the second is that it ignores their specific capacity to draw critically from science and use the resources it provides inventively to elaborate alternatives to disenchanted instrumentalism. To illustrate this, I look first at the discourse of two contemporary religious groups, one involving traditional belief, Christian creationism, and one new religious movement, Scientology. In both cases, enchanted versions of science are drawn out from critical engagement with the rhetoric of rationalized science, which far from displacing the charismatic relies heavily on this for its popular appeal.

Creating Design

In Chapter 3, I set out a number of discursive possibilities – or to take a slightly torn leaf from Habermas' book (1984, 1987), potentials for argumentation[1] – in addition to disenchantment opened up by a re-crafted view of rationalization. I associated disenchantment specifically with what Weber called 'formal rationality', referring especially to a combination of empiricist and rationalist discourses that provide the scientific community with major legitimating rhetorics of public self-presentation. In addition, however, I suggested there are three further dimensions to the discourse of modern science that Weber tended to downplay in his concern with what he considered to be the appropriate ethic of conduct for the scientist. These are: a discourse of formal practice or methodology that provides scientists with resources to dispute factual claims on the basis of experimental or other procedural adequacy ('formal irrationality'); 'substantive' discourses (both 'rational' and 'irrational') that provide scientists with means of accounting for error by reference to social and personal factors ('interests') intervening in the 'formal' activity of science ('the war of the gods'); and an additional discourse of public self-presentation and legitimation arising out of the distinction between two aspects of 'formal rationality' that in the modern world Weber treated as bound together. These are 'theoretical mastery' and 'calculation of means'. Whilst the

1 The leaf is torn as Habermas actually writes of '*structural* potentials for argumentation'.

latter is associated with instrumentalism, the former can be re-crafted through Whitehead's distinction between charisma as an attribute of individuals and as a dimension of experience characterized by increasing abstraction 'into the remote distance'. This enables and encourages a universalizing discourse of science that transcends instrumentalism, as seen in the grand moral vision (or 'metanarrative') of positivism. In effect, this is an enchanted discourse of science and, in keeping with Lessl's analysis of the 'priestly voice', such transcendent visions are crucial in scientists' formal, public self-presentation. However, the combination of these discursive aspects provide rhetorical resources for both scientists and non-scientists alike to envisage and articulate alternative cosmologies that incorporate science, not so much 'accommodatively', but in their fundamental constitution, as *discursive syncretisms* (Locke 1999a). This is what can be seen in the case of creationism at a number of levels.

'Creationism' is a category that has been used to refer to a range of beliefs (Coleman and Carlin 2004), but in parts of the developed world over recent decades is particularly associated with a form of fundamentalist Christianity, sometimes self-described as 'creation science', that has attracted considerable controversy because of its rejection of 'macro-evolution' – that is, evolutionary development across species (or 'kinds') – on what are claimed to be scientific grounds. The creationist movement is particularly prominent in the United States, where a division has been noted between groups that aim to pursue educational objectives through state policy and those confining themselves to scientific research (Toumey 1994). This division itself is of note in the light of the earlier discussion of rationalization, as it shows that the distinction between 'formal' and 'substantive' constructions may be mobilized to draw boundary demarcations within groups whose scientific status is itself contended. To turn this around somewhat, the distinction generates a dilemma for creationists over how best to pursue their goals. Given that the scientific status of their views is already contended by 'orthodox' scientists ('evolutionists'), they are faced with the problem that if they pursue activities that can be characterized as 'non-scientific', such as 'political' objectives involving educational policy, then they are potentially playing to their critics' charge that their *real* aims are not of a scientific nature. However, if they confine themselves to 'pure' science they risk losing public support from those whose primary concern is to uphold their religious beliefs against the perceived threat of evolution. Moreover, given that science is a major element in the curriculum of modern education – and in the view of some (for example, Nelkin 1982) was the proximate cause for the apparent resurgence of grassroots creationism in the United States in the late 1960s – not to address educational matters might seem perverse: what exactly is non-scientific about science education? Thus the distinction between 'formal' and 'substantive' is rhetorically charged: it is a means of constructing boundary demarcations, but the boundary is moveable as where 'science' ends and 'politics' or 'religion' begins is open to different construction and hence argument.

This is further shown by the legal challenges that have arisen in the United States following creationist involvement in state education. The most recent case centred specifically on the question of 'intelligent design' (ID) (Cole 2006, Edmond and Mercer 2006, Fuller 2006, 2007, Lambert 2006, Lynch 2006) and raises the issue of whether this is really only creationism by another name. Although I take no particular position on this debate here, it is the case that purported evidence of design in nature is a recurring feature of creationist discourse (Locke 1999a: 118-122) and I focus mainly on this in what follows.

The legal cases involving creationism pose a puzzle for the standard view of rationalization in which modernity is marked by progressively unfolding formal rationalization, such that all value spheres coalesce around the same central logic of instrumentalism, albeit ostensibly directed toward distinctive ends. Increasingly, however, even these are under threat from formally instrumental means of action becoming an end in itself, as in Weber's (1968) account of bureaucratization. This kind of view informs Habermas' (1984, 1987) notion of the 'colonization of the lifeworld' by technical, instrumental rationality. The implication is that the ostensibly distinct value spheres become oriented toward an identical modus operandi such that any differences between them are eroded producing a single, monolithic order – 'one-dimensionality' in Marcuse's (1964) expression. So in the case of law, for example, legal decision-making, however ostensibly oriented toward norms of communicatively established collective judgement through processes of argumentation between equally competent rational beings, would actually be increasingly characterized by the purely formal application of technically proscribed procedures that restrict capacities for communicative argumentation in a pre-defined fashion or simply by-pass them altogether. The end point of this would be the substitution of argumentation entirely by technical procedure. In substantial part, such substitution can be legitimized through appeal to science as the primary sphere of technical rationality; a case in point might be the increasing use of forensic science in courts as the basis of legal decision-making as in, for example, DNA-fingerprinting (Nelkin and Lindee 1995). In a word, law is scientized.

However, creationism casts some doubt on this type of view as here, to the contrary, science becomes legalized. The legal hearings regarding the teaching of creationism, whether through 'equal-time' with evolution or through ID, arise because of the particular feature of the United States Constitution that explicitly codifies a differentiation between church and state (Locke 2004b, Fuller 2006). Hence, state support for anything that can be defined as religious activity *in a manner adjudged to satisfy the demands of legal proof* is deemed unconstitutional and thus outlawed. An important feature of modern legal reasoning is that it formally institutionalizes the recognition that a case can be made for both sides of an argument (Billig 1996) even if it is held that there is ultimately only one truth to be had. That attaining this one truth is not necessarily so easy, however, is built into the machinery of legal proceedings in the existence of judges and juries. There is, in other words, a recognition that ultimately *someone* must decide where the truth rests between contending parties. Thus, although these legal-rational institutions may be attributed to a process of rationalization that differentiates value spheres on the basis of their formal logics, the extent to which their procedures are purely instrumentally formalizable is questionable; rather there is good reason to argue that the procedures of judgement remain those of the commonsense lifeworld characterized by what Pollner (1974, 1987) calls 'mundane reasoning'. Decision-making in any context can never be entirely formalized as it necessarily involves deciding the applicability of whichever rule is decided upon (Boden 1994) – and because of the rule about rules (see Chapter 1) this always involves an unspecified and unspecifiable network of background assumptions that in effect implicates a whole way of life with its associated cosmological beliefs. Moreover, in contexts such as those of courts of law where there are contending cases each of which can be made with equivalent credibility (until that is the moment of decision at which point the losing side becomes reconstructed as – and as always having been – *in*credible), the grounds for decision-making have a necessarily rhetorical character; they are themselves part of the argument. In the case of creationism this includes science. Thus, science is *part of the argument* not the means of its deciding even if it is presented as such after the event (Cambrosio, Keating and MacKenzie 1990, Lynch 1997).

This is clearly shown in cases involving creationism where the outcome has rested crucially on judges deciding that it is not science but really religion. The basis of these judgements has involved, amongst other things, drawing on established models of science as presented by the plaintiffs (Nelkin 1982, LaFollette 1983, Fuller 2006, 2007, Winiecki 2008). In other words, the cases have been decided by legal decisions over what constitutes science; thus, these are not cases of science deciding the law, but law deciding science. Moreover, it can be argued that the decisions were made on *substantively* and not formally rational grounds. It will be recalled from Chapter 3 that with respect to the law, Weber defines formal rationality as involving 'only unambiguous general characteristics of the facts of the case [being] taken into account' either as 'sense data' or 'through the logical analysis of meaning and where accordingly definitely fixed legal concepts in the form of highly abstract rules are formulated and applied'. Substantive rationality, on the other hand, involves legal decisions being 'influenced by norms different from those obtained through logical generalization of abstract interpretations of meaning ... includ[ing] ethical imperatives'. Now, it might be argued that the creationist cases have involved decisions of the former kind, in so far as judges have based them on the logical analysis of the meaning of 'science' – except precisely what the cases turn on *is* the meaning of 'science'. In other words, the judgements are not the outcome of simply formulating and applying fixed legal concepts through abstract rules to a given meaning of 'science'; they do involve this, but only *after* the meaning of 'science' has itself been decided.

For example, in the 1981 Arkansas case, Judge Overton's (1982: 202) judgement against the 'balanced treatment' teaching of creationism was explicit in stating that '[t]here is no controversy over the legal standards under which the Establishment Clause portion of this case must be judged', that is, the 'establishment of religion' clause of the First Amendment. In other words, 'definitely fixed legal concepts in the form of highly abstract rules' relevant to the case specifying the separation of church and state had already been formulated and were there to be applied. In order to apply them, however, Overton (1982: 217) had to establish the meaning of 'science' and for this he drew upon two kinds of definition offered by witnesses for the plaintiffs, one that stressed 'testab[ility] against the empirical world' (and more specifically still 'falsifiability') and another that stressed the 'accepted' actions and beliefs of the 'scientific community'. Thus, in part his judgement involved deciding that creationism breached the accepted actions and beliefs – the norms – of the scientific community (see also Winiecki 2008). So, having first decided that such a normative model of science could legitimately be used, he then employed it to help establish that creationism was really religion and thus that its teaching breached the establishment of religion clause. Arguably, therefore, 'balanced treatment' was overturned not on purely formal legal grounds, but at least partly on substantive grounds since the use of a normative model of science involves reference to 'ethical imperatives'. Moreover, there is a certain irony to this since creationists themselves draw upon both normative and empiricistic models of science in an occasioned fashion to construct their arguments against evolution (Locke 1999a: 71-99). Thus, all parties in the dispute – plaintiffs, defendents and judge – adopted similar argumentative strategies in which specific meanings of science were mobilized as standardized verbal formulations each to legitimize their actions and beliefs. Science, then, was used as a rhetorical resource not as a purely formal technical instrument; it only appeared as such *after* and specifically *as* the settlement of the legal argument.

It is not my intention, however, to argue that this legal ruling *was* 'substantively rational' as this would be to engage in a 'social interests' analysis (see Chapter 3). My point rather is that such an argument can be made and that this problematizes the distinction between 'formal' and 'substantive', such that this itself can become part of the argument (one person's '*Zweck*' is another's '*Wert*' and vice versa). This is especially clear in the matter of creationism, because it can be taken to provide support for the view that the 'war of the gods' continues within science itself. It may seem that there is a certain redundancy in stating this, but only if it is accepted that creationism is actually part of science – that it is heresy (Lessl 1988) rather than apostasy. But this is part of the argument: creationists claim creationism/ ID is scientific; their critics claim it is not – and both mobilize various models of science in support of their views. With respect to science, then, the 'war of the gods' is as much a dispute over whether there is a war to be fought as it is about who is winning – and the meaning of science is part of both armoury and spoils. In this, some creationists at least employ a similar discourse and range of rhetorical techniques as defenders of evolution (Lessl 1988, Prelli 1989a, Taylor 1992, 1996, Locke 1994, 1999a); in short, they argue like scientists even if it is not accepted by their opponents that this is sufficient to make them scientists. In terms of re-crafted rationalization, it shows either (or both) that, rather than simply displaying a formally rational procedural and analytic logic, science is characterized by a mixed discourse bearing both 'formal' and 'substantive' features that provides resources of internal argumentation, and rather than simply producing a widespread mental orientation of disenchantment, this discourse can itself help to provoke 'substantive' responses and developments. Thus, far from settling the war of the gods, science fuels it.

ID illustrates this in both ways. The creationist publications I have mainly studied, those of the British Creation Science Movement (CSM), often refer in support of ID to evidence provided by 'orthodox' scientific research that seems to show nature to be both enormously complicated and yet also extremely fine-tuned at least from a human point of view (for example, Rosevear 1991). This evidence is presented in the rhetoric of empiricism using the standard discursive form employed by scientists when describing features of the natural world (see Chapter 3). It appears, then, as purely objective description of the way the world is. This discourse itself provides a resource to legitimize the claims of ID, contributing to its 'factual' status and demonstrating the basic SSK view that empirical data is open to different interpretation (Collins 1983). Further, the existence of competing interpretations of nature within science also provides resources that can be used by 'non-scientists' to argue against specific scientific schools of thought and advance alternatives. So, for example, that there are different views amongst space scientists regarding the 'Principle of Mediocrity' (see Chapter 1) concerning the likelihood of other earth-like planets in the galaxy might itself be used as grounds for supporting an alternative like the 'anthropic principle', the view that the universe seems to have been designed with an earth-type planet and human beings in mind. A similar example concerns the debate between evolutionists over neo-Darwinian gradualism (Dawkins 2006) and punctuated equilibria (Eldredge 2000), which, although evolution per se is not in dispute, creationists have employed as a means to reject macro-evolution entirely (Lyne and Howe 1997; see also Park 2001). Further still, that scientists may themselves employ non-'formally rational' modes of argument in support of their views against opponents can lend additional resources to non-scientists. So, rather than disenchanting the world, science may however unintentionally lend strength to existing moral frameworks such as those of established religion. This is all the more so given that, as argued in Chapter 3, some scientists employ synecdoche in their public self-presentations and in so doing offer us enchanted images of the world. To explore these points further, I want to show how this works in one specific case involving the construction of what I call an enchanted self.

The Enchanted Self

To examine in detail the workings of synecdoche incorporating scientific knowledge claims in creationist discourse, I will closely scrutinize one specific example from a pamphlet published by CSM (Reeves 1999). It is important to appreciate that there is nothing particularly special about this pamphlet; it is just one of a regular bi-monthly series published by CSM which, so far as I am aware, has risen to no greater position of prominence than this. But it is just this apparent insignificance that is important – it is the ordinariness, the mundaneness (at least from a creationist point of view) that matters. It is by no means the only such example that could be used; as stated, the topic of design is a recurring one in these publications. However, this one is helpful for my current purpose because the workings of synecdoche are clear, especially in the way the reader is invited to consider themselves to be enchanted through the knowledge of nature disclosed by science. This is done through a focus on the functioning of the human body such that, while humans are reduced to bodily functioning, this is used to present us as evidencing God's created order, His intelligent design. Thus, an ostensibly disenchanted view of self as functional body is used to construct an enchanted view of self as part of God's mysterious and unknowable creation.

The discussion of the 'priestly voice' in Chapter 3 followed Lessl in arguing that this refers to a particular rhetorical style for managing the tension between in-group exclusiveness and out-group involvement, by at once legitimizing the prevailing social exclusivity through reference to the group's cosmological belief and inviting participation to those who accept the definition of humanity presented within this belief. The central rhetorical figure in this is synecdoche in which the part is defined by reference to the whole, such that a cosmic connection is forged involving the scientizing of the human even as science is humanized. Accordingly, individuals are invited to see themselves through and as a specific type of identity, an image of who and what they are as defined by the proffered belief that is at once a universal cosmology and a social universe. Thus, the priestly voice locates us in, through and as part of the order of reality it defines, encompassing both nature and society. The use of such enchanted imagery by 'orthodox' scientists provides opportunities for creationists to take up the same kind of rhetoric, but they do so selectively and in a manner that re-crafts it to suit their alternative worldview. This involves distancing themselves from any acceptance of macro-evolution, while also attempting to demonstrate coherence between the biblical Word and the world as presented in scientific descriptions. Thus, a key feature of creationist discourse involves attempting to show that the natural world as presented by science reflects the image of God as presented by creationists' reading of the Bible. Synecdochic constructions of the enchanted self play two key roles within this. First, the world is presented as there to be looked upon with an enchanted gaze, as an object of wonder and awe to be continuously amazed and astounded by as the remarkable phenomena of nature are displayed before the senses. In keeping with this, the world is objectified, treated as an independent, external, self-consistent and integral reality that can then be made the object of the enchanted self's regard: it is *there* for the enchanted self to wonder at.[2] Thus nature is made both a spectacle

2 In what I can only call a *wonderful* irony, similar rhetoric is employed by Richard Dawkins (1998). Compare also his (1998: 9) use of a metaphor of 'wheels' to describe the workings of cells with the discussion below. Thanks to Mike Hawkins for bringing this to my attention.

and spectacularized. Second, included within it as an object of wonder is the means whereby the enchanted self is able to regard the world enchantedly, that is the human body with its physical, sensory apparatus. This apparatus is also part of God's created order and so the enchantment of nature incorporates also the enchantment of the self-as-body. The enchanted self is then self-enchanting, regarding itself as an object of enchanted regard.

These features can be seen in the CSM pamphlet through its description and discussion of the functions of the enzyme ATP synthase, which makes specific reference to the work of a team of scientists under 'Dr. Walker'. To bring out the workings of the rhetoric of enchantment fully, I quote from the pamphlet at some length before presenting my analysis (as far as possible and unless otherwise noted, extracted quotations adopt the same spelling, punctuation, emphases and lay-out as the original text). The pamphlet opens with a brief comment about the importance of the wheel to human life as an introduction to a description of the enzyme, which then continues as follows (Reeves 1999: 1):

> miniature wheels are put to work in every cell of all plants and animals, and are responsible for generating the energy to enable them to function.
>
> **Superlatives**
>
> The wheel is found in the enzyme ATP synthase ... The wheel in this enzyme rotates at about one hundred revolutions per second. This miniature motor is 200,000 times smaller than a pinhead. Every cell in your body, and those of all living things, has hundreds, if not thousands of these motors. Your body contains over 10,000,000,000,000,000 of these motors.
>
> **What the motor does**
>
> The ATP synthase motor's job is to make the molecule adenosine triphosphate (ATP) from adenosine diphosphate (ADP) and phosphoric acid, a synthesis which requires an input of energy. The ATP can then break down to ADP again, giving up the energy by coupling itself to another chemical process within the cell which requires the energy in order to react. So energy is directed and the products are recycled - a neat scheme.
>
> As you read this, ATP is supplying the energy for the functioning of your brain, the beating of your heart and the contraction of your muscles. Similar to the release of energy stored in a compressed spring, so energy locked up in the ATP molecule, when triggered chemically, is released and made available to do work in our cells ...
>
> These extremely complex spinning motors are brilliantly designed. Each motor is built from thirty-one separate proteins that in turn are made from thousands of precisely arranged amino acids. ATP synthase is one of the most complex

> biological molecules ever pictured. Walker's team had been working for almost twenty years to elucidate the structure, atom by atom.

Following this, a more detailed description of the working of the enzyme is given making repeated references to it as a 'wheel' with associated imagery such as mention of a 'bent axle' and spinning 'motors'. The productive capacity of the enzyme is re-emphasized and the section concludes as follows (Reeves 1999: 2):

> All this to keep our bodies and brains functioning, thus making life possible ... As Dr. Walker comments, "It is incredible to think of these motors of life spinning around in our bodies!" Of course, the same amazing, ultra-miniature motors are spinning away in all living things, including plants, fungi and bacteria.

From these quotations, four features involved in the construction of an enchanted self can be drawn out: the objectification of nature; the artfulness of nature; the comparative artlessness of humans; and the artfulness of human bodies. Together, these define a synecdochic reduction in which human beings are viewed functionally and systemically as physical bodies within the natural world – and yet, contrary to what might be expected, nature (and we) are not thereby disenchanted but enchanted by virtue of nature (and us) being evidence of God's intelligent design and as such objects of awe and wonderment, 'fearfully and wonderfully made'. I will deal with these four features in turn.

The objectification of nature is a recurring feature of creationists' discourse just as it is in orthodox scientists' discourse. As discussed in Chapter 3, the adoption of the passive voice by scientists in formal, public presentations has been well documented (McCloskey 1983, Gilbert and Mulkay 1984, Mulkay 1985, Nelson, Megill, and McCloskey 1987, Bazerman 1981, 1988, Prelli 1989a, Myers 1990, 1997, Gross 1991, Halloran 1997, Hand and Velody 1997, Martin and Veel 1998, Zdenek 1998, Ziman 1968, 2000). As a feature of empiricist discourse, it works to obscure human agency in the process of knowledge production and invest it instead in the objects and phenomena presented in the text, which appear to take on a life of their own within the world described. Such a transformation appears in the description of ATP synthase through attributing a range of actions directly to the enzyme, such as 'coupling itself to another chemical process', 'supplying energy' and 'spinning away'. Moreover, it is said to be 'responsible', an attribute typically associated with intentionality and arguably evoking an ethic of conduct; indeed, the workings of this enzyme are said to 'make life possible' – truly a god-like ability! Similarly, alongside ATP synthase, various other kinds of objects and the relations between them are described as externalized, independent role-players in relation to which the enzyme is referred to as having a 'job', attributing to it a specifically human and social quality. Thus, the enzyme is transformed through metaphor into a dutiful worker, performing specific 'functions' as part of a wider system of energy production. Much might be made of this choice of metaphor in terms of the type of identity it draws on given the social context of modern societies.

The use of a 'factory' metaphor to describe aspects of the natural world (van Dijk 1998) might be seen as characteristic of the discourse of disenchantment, in which life is reduced to technicistic functionality (Weizenbaum 1984), but this is far from the whole story. For one thing, empiricist discourse, despite its construction of a world of dehumanized agents, nonetheless – and thereby – offers an identity to the reader. It invites us to regard the world precisely as an object consisting of an independent class of things that engage in various activities there to be looked upon; thus, it positions us as onlookers and spectators. This positioning is central to the discourses of both science and creationism, albeit that the particular directions in which they would have us look and the particular objects they would have us see are not always identically configured. A recurring feature of creationism in this respect is the instruction it gives us to look upon nature as artful.

The artfulness of nature can be seen in two dimensions of the above description: first, in references to the 'neatness' of the enzyme's functioning, its 'complexity', 'brilliant design' and 'superlative' quality; second, in the use of adjectives such as 'incredible' and 'amazing'. The first set of references purport to describe qualities inherent in nature; the second act as prescriptions or instructions as to how these qualities are to be regarded by us. Thus, the 'neatness' resides in the 'scheme' of energy production and recycling in the ATP-ADP reaction; the 'complexity', further enhanced by extrematization (Potter 1996) resides in the 'spinning motors' themselves, as does the 'design'. These are all qualities that the natural objects and processes themselves possess. The 'incredulity' and 'amazement' however are directions for us, instructions as to how to regard the natural phenomena being described. These two dimensions combine to produce a sense of the natural world as something that is both marvellous in and of itself, and worthy of being marvelled at; the fact that nature is artful – neat and complex and cleverly designed – warrants the instruction that it is to be viewed for its artfulness – as incredible and amazing. The quality of being superlative applies both to the object in itself and to the way it is to be viewed.

This sense of the superlative artfulness of nature is enhanced by an additional feature that serves to reinforce the invitation to adopt a specific image of self: the comparative artlessness of humans. This is seen here especially in the reference to the work of Walker's team. The sense conveyed in this brief description is of a long, painstaking task. This comes through a specification of the number of years of work involved as 'almost twenty' and a description of the nature of the work as 'elucidation atom by atom', giving an overall impression of great difficulty. The way time periods are specified in discourse carries significance, not only in the length of time referred to but the precise formulation in context. The time period 'twenty years' may mean different things in different contexts, notably it may or may not mean a relatively long time. To speak of twenty years in respect of, say, the early history of the solar system is a relatively short time (even for Young

Earthers[3]), but in reference to an individual person's working life it is so much longer. Moreover, in saying '*almost* twenty years', emphasis is placed on the end of a specified period, rounding the period of time up and so accentuating its length. Thus, 'almost twenty years' here equates to a substantial portion of a person's working life. Further, the specification of the nature of the work as 'atom by atom' makes it into a difficult and painstaking task. Atoms are known, even by much of the public (Evans and Durant 1989) to be extremely small; thus whether or not one takes the description of the work literally, the sense is conveyed of an immensely laborious and meticulous task. Yet what appears to have been such a difficult task for human workers is evidently merely a matter of routine for nature, which is producing ATP synthase molecules in the billions of billions, in 'every cell of all living things' as a matter of daily course. Thus, the stress on the difficulty of the human labour serves to enhance the sense that nature is far more clever, complex and artful than we are. Human toil and ingenuity pales before such complexity, which is accomplished without us and before us – and indeed is necessary for us even to have begun to undertake the task of its unravelling, a task that required tens of thousands of human worker-hours. This sense of nature's artfulness is further enhanced by the root metaphor of the wheel, a point to which I return shortly.

However, as much as our labours may be put into the shade by such natural 'brilliance', our bodies are made that much more the object of our amazement. The quotation is especially explicit in this regard in adopting a voice that speaks to the reader directly as 'you'. With this, the text invites us as we read to consider the workings of our own bodies. We are each individually explicitly included in the phenomena being described. At the same time, however, it is our bodies, their organs and functioning that we are invited to consider as independent, external systems. However, this is not disenchantment, but the very opposite as it enhances the invitation to be amazed: all this 'incredible' activity is going on inside you even as you read; all this energy production activity involving these billions of tiny, but dauntingly complex wheels, spinning away, on and on, without you and yet absolutely with you, because without them you would not be here and none of this reading would be happening at all. Thus, although readers are invited to view themselves as their bodies and to view their bodies as functioning systems, it is this very functioning that is enchanted. Our bodies are presented to us as artful and as part of the incredibly complex artfulness of nature that precedes us and in relation to which our own activity is comparatively dull and laborious. It is significant here that Walker's team is described as 'elucidating' nature. To elucidate is to make clear something that already exists: the complexity goes before us (but not our bodies); all we can hope to do for ourselves is bring out the light that already exists in nature (our bodies). Thus, the image of self we are offered is one that looks on nature as a wondrous spectacle and holds the human body in higher regard than

3 'Young Earther' is the term used by creationists to refer to those who believe that the creation week recorded in *Genesis* occurred some six to ten thousand years ago.

human labour as an object that partakes of the enchanted quality of the natural world.

The final move in this synecdochic reduction of self to enchanted body is the connecting of the natural world to God. Thus, after a section arguing against the adequacy of an evolutionary explanation of the enzyme, the pamphlet concludes as follows (Reeves 1999: 4):

> **Who did design the wheel?**
>
> We have noted that the wheel [that is, the enzyme] is present in all forms of life. It is part of a motor that is irreducibly complicated. Its proteins are coded for by information on genes. Information Theory tells us that information is corrupted by chance processes. Information only derives from an intelligent source. The genetic information is translated and the motor manufactured and assembled by a series of mechanisms which are, in total, even more irreducibly complex than the motor itself. (A pin-making machine is more complex than the pin it produces!) The Designer of the wheel perforce designed the whole life process.
>
> Man also re-invented the wheel, without realising that he bore within himself some ten quadrillion tiny whirling wheels. Man's motors are, by comparison, simple, even clumsy, and usually noisy. This enzyme is nice in both its meticulous design and its beneficial effect.
>
> The psalmist David (139:14) declares 'I will praise Thee; for I am fearfully and wonderfully made: marvellous are Thy works; and that my soul knoweth right well.' ... Awesome ... just about sums it up. The more that is discovered of the Creator's ways, the more we stand in awe of Him. Ultimately, His ways are past finding out.

There are a number of knowledge-claims and argumentative features in this section the validity of which are not at issue here. What I wish to draw attention to is the manner in which the passage invites us not only to see the presence of God, 'the Designer' in the enzyme, but to see ourselves as 'man' (that is, humanity) in relation to Him. As already stated, humanity appears here as comparatively inadequate not only in the particular qualities that human 'motors' have, but also in that we have merely 're-invented' something already designed by God. This echoes the introductory section of the pamphlet, where the commonplace notion of 're-inventing the wheel' is referred to as describing 'unnecessary research' (Reeves 1999: 1). Thus, human labour and technology are similarly unnecessary, poor imitations at best of the 'irreducibly complex' natural systems made by God. Moreover, His makings are 'ultimately past finding out', beyond our ability even to know let alone replicate or better. As is clear from the biblical quotation, these makings include us: the Psalmist is here given the voice to speak on behalf of us all – his 'I' becomes the 'we' who 'stand in awe of the Creator's ways', including our own bodies.

Thus, we and the world are enchanted by and through science, the very labour of which accentuates the marvel and wonder of nature made by God beyond our ken. Science, then, does not necessarily disenchant – just the opposite: the doing of science is open to being represented as making available the wonder of the world and a world to be wondered at. Contrary to the standard version of rationalization, enchanted discourse remains a significant feature of scientists' rhetoric. Indeed, the very claim to provide definitive knowledge of the world is itself a component of this, because the world it promises to disclose can be presented as a world of wonders filled with as yet unknown complexity, with secrets to be found tortuously involving difficult and extensive human labour. Whilst such representations are open to different interpretation, they can still be taken up by ostensibly non-scientific others like creationists to demonstrate the awesomeness of God: the more of nature that becomes known, the more our sense of knowing is beggared. In inviting people to wonder at nature, including ourselves as part of the natural order God made, creationists hold out for us this sense of enchantment and they are even able to use the very materialist reductionism of science as a resource to this end.

Rationalization, therefore, should not be equated with such reductionism and nor with disenchantment. What needs to be recognized is that the meaning of science – both in general and in the specifics of its knowledge-claims, procedures, actions and beliefs – is not simply given by 'formal' definition. Rather, it is something that is continuously being worked out and, in various ways and to varying degrees, this involves us all regardless of whether or not we are considered or consider ourselves to be scientists. A re-crafted rationalization accepts this, because it treats the meaning of science not as in-built or pre-given, but as a matter of ongoing social, public argumentation – argumentation that occurs in the everyday lifeworld as much as in specialized institutions such as courts of law, whose discourse in any case articulates around the same reflexive constituents as mundane reasoning. Thus, whether science disenchants or not, whether it instrumentalizes or not and even whether it secularizes or not are questions for us all, at least potentially, without necessary or pre-determined answers. Rationalization poses us these questions and accordingly we find in the social world versions of science that are disenchanted and versions that are enchanted.

ID is one such enchanted version. For creationists it is one they can develop and use to advance their 'old-time' religion; for others it may be used to different ends – perhaps even as a revitalized metanarrative for science in its quest to 'know the mind of God' (Fuller 2007). But if so, this would no more necessarily mean disenchantment than did positivism; rather, it could equally well be seen as a further stage in the abstraction of the charismatic 'into the remote distance' as envisioned by Whitehead. Moreover, ID is by no means the only contemporary form of such an abstracted charisma. As Whitehead recognized, other forms are to be found in some new religious movements such as Scientology. Here again, an enchanted image of self is drawn in part from science and it is to a consideration of this that I turn next.

Chapter 5
Soulful Science – Scientology

In Chapter 4, I considered a case of a traditionally-based religious movement, Christian creationism to argue that its selective, critical and interpretative use of science could be understood in terms of the re-crafted view of rationalization set out in Part I. In this chapter, I look at a contrasting type of contemporary religion, Scientology to argue a similar case. As with creationism, I aim to show that Scientology offers an enchanted self that is important to its appeal and is built in substantial part from resources offered by modern science. However, as with creationism, the discourse of Scientology is informed by a critical view of science; thus, it is not simply a direct articulation of a scientized, rationalized outlook, but a selective and critically re-crafted version. This is made possible in part by the public availability of a mixed scientific discourse, but is also constructed from resources provided by alternative sources of knowledge and belief. Just as creationists employ materials from the Bible to construct their critical re-crafting of science, so do Scientologists draw on materials from ostensibly non-scientific sources, which include a wide range of beliefs from diverse cultures (Kent 1996). These are presented as having the same essential concern to which Scientology provides the answer and as such it is the unifying culmination of human traditions of thought. Central to this claim is the rhetorical figure of synecdoche and in this Scientologists intone the same priestly incantation as some scientists and some creationists. Further, in its bid to unite both western and eastern forms of belief, Scientology is comparable to a number of other religious movements founded in the twentieth century bearing certain 'New Age' characteristics (York 1995, Heelas 1996, Hanegraaff 1999; see also Main 1999). Of central importance amongst these are discursive features that have been taken to provide grounds for doubting the standard view of rationalization through the notion of 'reflexive spirituality' (Besecke 2001). This has much in common with my own analysis, although there is a critical point of difference to be discussed before looking at Scientology in more detail.

Reflexive Spirituality

Besecke's discussion is of considerable pertinence, because it strikes a resonant note with Whitehead's view of the further abstraction of charisma. As discussed in Chapter 3, Whitehead anticipated a movement away from attachment to specific 'concrete symbolism' towards interest in the nature of symbolization itself. She did not go into detail, but one of the things this could involve is a shift from attaching specific traditional meanings to religious symbols or treating them in a 'literal' manner towards interpreting them as representations of something more general and transcendent. In so doing, the sense of direct, totemic connection between the symbol and a particular social group with an associated ethic of conduct and established set of ritual practices would be broken, enabling possible connections to be forged with alternative ways of life and their associated symbols. Or it might entail a concern with practice as such, not in the instrumental and technical sense bereft of attachment to ultimate meaning that concerned Weber, but rather with attention given to the quality of the doing and how this manifests or is informed by the generalized transcendent understanding taken to be expressed by symbolization. Most importantly, it would be claimed that this transcendent understanding encompasses and informs both religion and science.

Something of these characteristics can be seen in the features Besecke identifies in the language of reflexive spirituality. Most significantly, she stresses that it articulates a dual commitment to belief in transcendent meaning and to upholding the principles of rationality associated with modern science. Thus, effort is made to acknowledge and accept the demands of modern reason, but it is not accepted that this necessarily undermines or invalidates spiritual meaning; instead, a transcendent meaning built from both reason and spirit, science and religion is sought. Besecke (2001: 372-376) identifies a number of specific discursive features that display this dual commitment, including 'metaphorical interpretation', 'pluralism', 'reflexivity' and 'mysticism'. Thus, 'metaphorical interpretation' involves treating religious beliefs that ostensibly conflict with science as symbolic rather than literal truths and interpreting them accordingly to fit in with scientific knowledge or understanding. 'Pluralism' involves drawing on a range of beliefs and traditions in the 'spiritual marketplace' and, together with science, creating a synthesis in a manner that Besecke (2001: 374) likens to 'the rational meaning-making techniques in academic work'. Similarly, 'reflexivity' involves viewing beliefs with some criticality in a context of multiple possibilities and making revisions in the light of 'expert' knowledge. Again, whilst such use of specialist sources 'affirm[s] the rational enterprise of academic scholarship', it does so she (2001: 376) argues 'as a method of attaining transcendent meaning'. Thus, techniques of rational meaning-making are employed but not instrumentally either as an end in themselves or solely to further the production of this-worldly knowledge; rather they are used towards the end of constructing a mode of understanding that transcends both science and established beliefs.

Now, it might seem a bit of an over-stretch to suggest that Whitehead had these kinds of features in mind, but a much clearer similarity of view is discernible between her description of the relationship between positivism and 'occultism', and Besecke's notion of 'mysticism'. By this, Besecke (2001: 373) means a feature of the discourse of reflexive spirituality that gives priority to a form of direct, personal experience of the transcendent which, despite its apparent subjectivity, 'seems to share with scientific rationality a value for empirical observation'. In not dissimilar fashion, Whitehead (1974: 564) argued that occultism, despite rejecting the materialist assumptions informing positivism, nonetheless shares an interest in the practical knowledge and control exhibited by science through technology. More significantly, occultism is also interested in the 'supra-sensible powers and states of mind to which the mystic alludes' and seeks an encompassing comprehension in which the apparent diversity of knowledges meet in a more fundamental unity. As Whitehead (1974: 565) put it: 'The "contentless comprehension" of the mystic is on some level the same as the cognitive control of the intellect which is again on some level related to the orderliness and regularity of nature's most solid and material processes.' Thus, for both what Besecke refers to as 'mysticism' and Whitehead as 'occultism', there is a similar stress on the value of empirical experience and practice, but neither sees this as representing a purely instrumental outlook since it is not pursued as an end in itself, but as a ground of validation and/or expression of a transcending comprehension. In part, the measure of its transcendency is precisely that it incorporates and enables practical efficacy.

Thus, despite some differences in choice and usage of terms, Besecke and Whitehead – in arguments separated by close to thirty years – seem to be pointing to similar kinds of characteristics in contemporary 'occulture' (Partridge 2004). Crucially, both emphasize what they see as a sense of renewed visionary transcendence that seeks neither simply to resist nor to accommodate science, but to incorporate it within a wider spiritual discourse. Accordingly, both reject the standard view of rationalization. It is important to stress this, because the period separating their accounts has been one in which this view has become consolidated within a range of sociological theorizing as discussed in Chapter 2. Similarly, the associated view of secularization as centrally involving a process of scientized disenchantment has become part of the standard definition used in introductory textbooks (for example, Fulcher and Scott 2007: 414-416; see also Davie 2007). The partiality of this view is clearly brought out by Besecke, just as it was in her rather different manner by Whitehead so much earlier. As Besecke points out, much sociological discussion of modernity tends to ignore the discourse of contemporary religion. There is considerable irony to this since, in expressing concern over the supposed meaninglessness wrought by disenchantment, theorists often point to the need for just the kind of moral discourse that she finds under development in reflexive spirituality.

A point to stress here is that reflexive spirituality is not properly understood if it is viewed as an individualistic religious impulse solely concerned with personal spiritual development in contrast to the collective orientation of traditional religion – it is not, Besecke (2001: 368) insists, 'Sheilaism'. Rather, it is 'public, communicative, interactive and social', as is evident in the fact that it takes a discursive form which necessarily requires 'a relationship to societal rationality'. In other words, even if it were the case that the contemporary religious sensibility is chiefly a consequence of fragmented, privatized individualism, this would still not provide a sufficient sociological account which has to attend to the *social* means whereby such a self-understanding can be both formed and *articulated*. This is why the focus on discourse is crucial: even if in the course of sitting and contemplating my navel, I experience some form of personal transcendent revelation, some 'contentless comprehension' of the grand unified theory of everything – be it God, ID or Richard Dawkins – any attempt to communicate this experience to others would necessitate making use of the available resources of representation and whether discursive, visual (Flanagan 2007, Flory and Miller 2007, Locke 2008) or whatever else, these are necessarily collectively held. Even for Plato, Truth (grandly unified with the Good and the Beautiful) existed only in the divine realm of Ideas and therefore necessarily required rhetoric – the available means of persuasion – as a 'handmaiden' for its public, oratorical representation (McGee and Lyne 1987, Zappen 1994). No matter how high their mountaintop, the gods must always come down to earth and be sullied by the vernacular if their voice is to be heard – precisely Lessl's point about science as discussed in Chapter 3.

The link to rhetoric can be developed to provide a different take on the contemporary religious 'supermarket'. This metaphor has been used in support of rationalization as evidence not just of the decline of an established unifying Church, but also of a modern outlook that treats beliefs instrumentally as consumer goods (Wilson 1990, Blasi 2009). However, although individuals in the modern world may be enabled to pick and choose their own variety pack of beliefs, two things are minimally required for them to do so: first, they have to be willing to take a trip to the store; and second, they have to be able to fit the various 'goods' together in a 'consumable' way. The 'supermarket' metaphor – like all metaphors (Lakoff and Johnson 1980) – tells at best only part of the story, that there is a storehouse of ideas available. But it does not capture the active work involved in going to get those ideas and putting them together. Why in a disenchanted world would people want to visit the spiritual supermarket at all?[1] One argument here is that these beliefs serve a therapeutic purpose essentially directed at practical efficacy, an argument to which I return shortly. But even if this is the primary motivation, it is not simply the case that people pick and choose their beliefs from a shelf of pre-packaged products; rather, they actively engage in constructing new combinations. This is amongst the things that, to reprise the argument made in Chapter 1, folks are doing. The products in the marketplace of beliefs may be available in pre-packaged varieties, but they can also be picked up as a range of separate ingredients in need of some 'cooking' to be made into something more 'edible' – and even the pre-packaged stuff does not come with full instructions for use in all circumstances. Ideas and understandings always have to be *put* into practice. Actions do not automatically follow from given 'recipes' or sets of rules, because the rules require interpretation to be made to apply in given circumstances

1 A similar question can be posed of the view that 'popular spiritualities' (Hume and McPhillips 2006) are characterized by instrumental rhetoric reflecting the prevailing 'habitus' of modern culture (Sutcliffe 2006): if this is so dominated by instrumentalism, why would people construct spiritual beliefs at all and how would they obtain the means to do so? This view seems also to assume an impoverished notion of rhetoric that associates it solely with instrumental motivations; suffice it to say that such a view is itself rhetorical.

(Garfinkel 1967). So, while there may be a wide range of alternative beliefs available in modern society, those beliefs still have to be worked to give them specific form and meaning in a manner that can be made relevant and applicable to contemporary and immediate circumstances. Beliefs have to be worked out, worked up and worked into specific moments and contexts of action; in a word, they have to be invented.

'Invention' is a word favoured by rhetoricians (Perelman and Olbrechts-Tyteca 1969, Nelson 1987, Billig 1996) to capture the sense that arguments must be relevant to the situation in which they are advanced and made specific to current circumstances (Jasinski 1997). In the classical tradition, rhetoric referred specifically to oratory, techniques of speechifying that might be employed to move an audience and persuade them towards one view against another (Gaonkar 1997). Such speech-making always had to be situational, using appropriate techniques and arguments to move the audience in an effective manner (Waddell 1997). Accordingly, although rhetoricians may have sought to document the range and forms of the means of persuasion, no general guidebook of rhetorical tricks of the trade could ever do the specific work required of the rhetor to make arguments relevant to the here and now. Rhetors have to be creative; they must invent – and when it comes to inventing cosmologies in the modern context, this can involve drawing on scientific discourse as much as that of any other belief. This is not rationalization in the standard sense of disenchantment; it is rather using the possibilities for different imagining that science has opened up even as it is itself an expression of such opening. As Besecke (2001: 368) puts it, the 'language of reflexive spirituality is primarily a method of making religious traditions meaningful for a rationalized social context' – with the proviso however that rationalization does not equate to instrumentalism.

She makes this point specifically in critique of the view that reflexive spirituality is no more than an alternative form of therapy. This is especially apposite to the present discussion because just this has been said of the particular case of Scientology (Wilson 1990). There is an irony about this, because the founder of Scientology, L. Ron Hubbard was explicit in rejecting such a view and was also strongly critical of the established institutions of personal therapy and their claims to scientific status (Wallis 1976: 234-236). This gels with Besecke's (2001: 376) further point that the language of reflexive spirituality enables members to 'critique the limits of rationality, without wholly rejecting rationality'. This, she argues, points to the need for a revised view of 'the relationship between religion and rationalization' as 'dynamic' such that 'religion can affect the process of rationalization at the same time as religion is itself affected by rationalization'. I made a similar point in Chapter 4 regarding creationism. However as I also suggested, to understand this dynamic properly, we need to look at how religion and science are mutually shaped into compatible forms and consider what this means for science as much as religion. In this respect, Besecke's analysis is one-sided as she tends to treat science and rationality in an unproblematic fashion as a kind of given base-line against which the language of spirituality shifts. She does not then give attention to how science itself is re-worked in accord with the dynamic process she advocates. For example, in looking at 'metaphorical interpretation', we should consider not only how traditional religious teachings are interpreted symbolically, but also how science is used metaphorically – as was seen in Chapter 4 in the creationist analogy between ATP synthase and 'the wheel' (see also Locke 1999a, Locke 2001a).[2] Similarly, whilst it is valuable to point to the similarities with academic reasoning, we should also consider how this is informed by religious ways of thinking and arguing, as in Lessl's analogy with the 'priestly voice' in popular science (see also Campbell 2001, Mellor 2003). In general then, if we are fully to grasp the dynamics involved in the process of rationalization and re-craft our understanding accordingly, we need to see not only how spiritual discourse is shaped to fit features associated with modern science, but also how science is shaped to fit spiritual discourse. We need to consider how the two are brought together into a mutually informing single cosmological vision, a discursive syncretism in which both disenchanted and enchanted versions of science are worked together. It is this that I now look at in the case of Scientology.

Scientology

Scientology is a good case in point because it has been subjected to two strongly contrasting sociological interpretations by Wilson (1990) and Whitehead (1974, 1987). As already mentioned, for Whitehead, Scientology is an example of contemporary 'occultism' that to some extent articulates a more abstracted level of the charismatic. In the terms I am using here, it is informed by an enchanted view of science. For Wilson on the other hand, it is an example of a 'manipulationist sect' with characteristics entirely understandable in terms of the standard view of rationalization as a product of a disenchanted worldview. Since Wilson's is the more established view in relation to wider sociological discussion and since the purpose of this book is to argue against this, I will devote my main attention to his account.[3] First, I will outline a brief history and description of the beliefs of Scientologists intended to draw out particular features of relevance to the issue of rationalization. This is deliberately presented in empiricist-descriptive terms to make the supposedly rationalized features of Scientology as sharp and apparently unproblematic as possible, the better then to show why this is insufficient. Needless to say, the description is not intended to be taken in any sense as definitive; it is constructed for a specific purpose and with rhetorical intent in keeping with rhetorical sociology's reflexive acknowledgement of its own inevitably ironic

---

2 Restivo (1978, 1982) provides a comparable analysis of Fritjof Capra's writings. For an example of the use of religious metaphors in science, see Nelkin and Lindee (1995) and van Dijck (1998) on the gene as the 'Book of Life'.

3 Wallis (1976) is justly the better known treatment but he adopts Wilson's (1970) categorization. For fuller discussion of all three views see Locke (2004a).

character (viewed as an anarchic strength – see Locke 2007 and Chapter 10). It is culled from a range of primary and secondary sources using the 'rational meaning-making techniques in academic work' – though quite how 'rational' they actually are is rather moot under the circumstances! Main primary sources include Hubbard (1987, 1996), Church of Scientology International (1998a, 1998b) and Citizen's Commission on Human Rights (CCHR 2000), and, in addition to the sociological accounts already mentioned, main secondary sources include Gardner (1957), Miller (1987), Berger (1989), Atack (1990) and Kent (1997, 1999). A variety of websites were also viewed. I have largely avoided giving specific references except to direct citations.

The Church of Scientology was founded by L. Ron Hubbard in 1954. It grew out of an alternative therapy called Dianetics that Hubbard had introduced six years earlier. Prior to this he had pursued a variety of careers including that of science fiction writer and had published a number of stories in the magazine, *Astounding Science Fiction*, then edited by John Campbell Jr, who helped in the development of Dianetics partly by publishing Hubbard's first essay outlining the therapy. This was followed by a highly successful book first published in 1950 (Hubbard 1987) that attracted international interest and following. The central therapeutic technique of Dianetics involved an activity Hubbard called 'auditing'. Wallis (1976: 31-38) likens this to psychoanalysis though Hubbard disputed such parallels, voicing strong criticism of established psychiatry that the Church continued after his death (for example, CCHR 2000; see also Kent 1999). Descriptions of auditing suggest it does involve close questioning about past experiences particularly involving traumatic events. The novitiate's reactions are registered on a machine called an 'e-meter' (short for 'electro-psychometer') specifically developed by Hubbard for the auditing process. It has been likened to a lie-detector (Atack 1990: 128), according to Kent (1999: 163, n.22) even by Hubbard himself and a fascinating account of the experience of auditing is given by Whitehead (1987). The purpose of auditing is to identify 'engrams', mental blocks produced by past traumas, the removal of which produces a state of 'clear'. In his description of the features of the 'analytic mind' that results from auditing, Hubbard (1987) provides an early example of a computer metaphor for the human mind, a form of 'metaphorical interpretation' that has since become much more common. Thus, a 'clear' can store and access all data at will, compute perfectly, learn swiftly with full comprehension and generally be in complete control of themselves and their actions. The analytic mind is contrasted to the 'reactive mind' where engrams are found and which is keyed automatically when stimulated by events similar to those that produced the original trauma. As the name suggests, the reactive mind is a stimulus-response mechanism and the aim of auditing is to replace its behaviouristic responses with the awareness and self-control of the analytic mind.

One point of note is that there is a recurring emphasis on the 'workability' of the therapy and its empirical grounding in audited cases, which might then be taken as evidence of disenchantment. Such a view is made potentially problematic, however, by Hubbard's next career move. Despite, or as Wallis (1976) argues, because of its popular success, the Dianetics movement soon fragmented prompting Hubbard to break away and found Scientology. This was billed as a religion rather than (just) a therapy, although it was said to have developed out of Dianetics and continued to use auditing as its central practice. Scientology went beyond therapy in advancing a full blown soteriology based around a view of humans as 'thetans'. Thetans are souls or spirits (Hubbard 1996: 30), defined as such against the material level of existence which includes both our bodies and our minds. They are immortal, all-powerful beings said to be trillions of years old and to have lived numerous lives throughout history and across space. This record of past lives is known as the 'Track' and is said to have become apparent from auditing, during which many individuals seemed to be recounting memories of previous existences some of which did not conform to the known record of human history while others were overtly extra-terrestrial. This led Hubbard (1996: 52) to propose that some engrams were inherited from traumas experienced in past lives. Thus, the process of auditing was extended so that even the state of 'clear' was seen as being relatively low-level compared to that of 'OT' (Operating Thetan). OTs are 'at cause' and able to create 'MEST' – matter, energy, space, and time; they can also 'exteriorize' – leave their bodies – and may also possess other paranormal abilities.

'Mystery to Formulae'

It is apparent even from this brief outline that Scientology draws heavily on certain kinds of representation of science, some aspects of which I return to in more detail below. Moreover, in the style of its ministration through the practice of auditing using the 'e-meter', as well as the practical, this-worldly goals the Church offers its members, it can readily be seen how Scientology might invite interpretation in terms of the standard view of rationalization, an invitation fully taken up by Wilson (1990). For Wilson, Scientology is a type of religious system that reflects a rationalized and secularized social environment marked by a number of defining characteristics the most important of which are: a pragmatic, utilitarian and instrumental emphasis on 'religion as an agency of progress and self-improvement' (Wilson 1990: 269) especially in therapeutic forms; a widening knowledge of other faiths and beliefs that relativizes religious truth; and the rise of the natural sciences that Wilson associates with an emphasis on technical specification, and standardization and routinization of procedures with the aim of achieving replicability of outcome – in short, Weber's 'calculation of means' as discussed in Chapter 3.

In Scientology, this is apparent in both its means and ends: its practice of ministration and its form of salvation. Auditing is characterized by 'precise specification' (Wilson 1990: 273) of procedure, involving techniques that are presented as replicable, standardized routines with predictable outcomes. They are 'technical devices … to increase the production of salvation' and as such have 'rationalized the path' to this end. Thus, as Wilson (1990: 274) neatly puts it, Hubbard reduced 'mystery to formulae'. Meanwhile, the form of salvation itself is directed towards individual improvement and achievement of potential in a practical and therapeutic form reflecting a rationalized, this-worldly orientation. Similarly, Scientology's basic ethic is directed at individual therapy through the achievement of self-determination and self-responsibility in keeping with the individualized ethos of modern secularized society. For Wilson (1990: 273), this constitutes the real appeal of Scientology for its followers; they are motivated to join not by its 'elaborate metaphysical system' but by its 'promise of personal therapy'. Thus, he (1990: 275) argues, there is 'a conjunction of technical means and spiritual goals', which can also be seen in the terminology used by the religion. For Wilson, it is no more than to be expected from the transformations wrought by secularization that religious expression should adopt 'new language' drawn from a therapeutic context. In accord with this, Scientology has sought 'clinically' to remove 'all non-neutral connotations' from its terminology; auditors, for example, do not rely on their own 'spiritual apprehensions or ... personal appraisal', but use 'prescribed procedures' referred to as standard 'tech'. In this, Wilson argues, the religion shows itself to be 'explicitly committed to the ideal of rational thought and self-examination' and thus 'bears the imprint of the technological age in which it came into existence'.

I have followed Wilson's description closely to bring out not only the thorough quality of his application of the standard view of rationalization, which on the face of it is very persuasive, but also the thoroughly rationalized character he ascribes to Scientology. It can clearly be seen from this that the notion of rationalization employed is that of a progressively advancing technical logic with an essentially instrumental and this-worldly orientation, the driving momentum of which comes from science and technology, which has stamped the surrounding society in its mould and provides the model on which the religious system is based. The religion appeals to a set of logics taken to be characteristic of science not only in its emphasis on instrumental goals, but also in the specific methodology for achieving those goals through replicable and technologically specifiable techniques of an emotionally-neutral form. The upshot of this is the displacement of 'mystery by formulae' – as pithy a definition of disenchantment as one could ask for.

However, in relation to the foregoing discussion, there are two particular features of Wilson's account that become more problematic with closer attention: his characterization of the language of Scientology; and his view of the motivations of its followers. It is clear that his view of its language is at odds with the description of the discourse of reflexive spirituality provided by Besecke. Two possible inferences might then be drawn, one of which is that Scientology is not a case of 'reflexive spirituality'. In support of this, there is little in Besecke's (2001) description of the practices of the group she focuses on, Common Ground, to suggest it shares the kinds of characteristics taken by Wilson to be critical about Scientology, such as routinized, technologized ministration specified by standard 'tech'. Nonetheless, significant similarities between the discourse of reflexive spirituality and the 'terminology' instituted by Hubbard can be found. These are sufficient in my view to justify the second inference, which is that Wilson's description of Scientology is flawed. More correctly, it is itself an example of the application of a discourse, the discourse of disenchantment that characterizes the standard sociological view of rationalization. The point here is not that Scientology cannot be read through this discourse and thus interpreted as evidence of such rationalization – Wilson's account clearly shows that it can; it is rather that such an interpretation has a rhetorical character, one that selectively emphasizes some features over others and interpretatively characterizes them in its own 'standardized' manner in instrumental terms. As such, it ignores other features of Hubbard's 'new language' and ways of interpreting this. Specifically, it ignores how Hubbard himself represents the 'spiritual goals' Scientology is directed towards and indeed that these 'goals' are explicitly *spiritual*, offering a form of salvation and a theodicy with a theory of suffering. Moreover, to describe the belief as purely 'this-worldly' seems more than a little at odds with its philosophical anthropology constructed around the notion of 'thetans', other-worldly entities of a particularly radical form responsible for the creation of 'this world' (as MEST) and which constitute our essential being whether we are aware of it or not. These are explicitly described as 'souls', a term Hubbard (1996) distinguishes from our material form in 'this world' including both mind and body. If this is (just) therapy, it is of an extraordinarily ambitious kind that one might be forgiven for thinking is of little practical utility if not deeply unrealistic.

However, of greater significance is that Hubbard's soteriology incorporates a critique of science in the very terms Wilson uses to characterize it, a critique concerned to argue that this is precisely *not* real 'science'. Hubbard is explicit in rejecting the instrumentalist and materialist, this-worldly version of science that he associates with the modern scientific worldview and which Wilson himself articulates. But in a crucial turnabout Hubbard does so *in the name of science itself*, which he seeks to reconstruct in a manner that transcends materialism and instrumentalism. Wilson both ignores this and, in using his own disenchanted version of science as the crucial tool of interpretation, effectively turns Hubbard into a shadow version of who he presents himself to be.

This is also highly significant to how we view Scientologists' motivations. Wilson's claim that Scientologists are motivated by 'the promise of personal therapy' is disputable from Wallis' (1976: 166-171) data that suggest a more mixed set of motives (see also Locke 2004a). Aside from this, however, the most compelling reason for disputing Wilson's view is his effective dismissal of the importance to Scientologists of Hubbard's 'elaborate metaphysical system'. It is not so much that many Scientologists might find this less persuasive than the offer of straightening out their personal lives, but the fact that it is presented *and can be presented at all* as justification and raison d'être for any therapeutic goals needs sociological understanding. The existence of this level and form of discourse must be included in any sociological account for it to be considered adequate. How is it possible and why does Hubbard bother with it? If the world is so disenchanted, if we are all so fully in the grip of an outlook of mind that is solely interested in means-to-ends manipulation of our immediate material circumstance, then what point is served by what can only seem an utterly fantastic, indeed ludicrously silly notion that we are actually cosmically powered beings who have simply forgotten about it – like you do, at least if you inhabit the plot of an average superhero comic. Why would anybody take this seriously? What social conditions must prevail for some at least to take it seriously and for someone to present it to them as a serious proposition? If we are to avoid falling back on the P.T. Barnum postulate (see Chapter 1), then we must assume there is more going on here than mere duplicity and we must also recognize that, whatever these social conditions are, they are not adequately described by disenchantment alone. For all his adoption of the terminology of instrumental technique, it needs also to be recognized and understood that Hubbard invites his followers into an enchanted world, one that re-crafts disenchanted science through a promise of transcendence that reaches beyond its perceived limitations.

Thus, like reflexive spirituality, Hubbard depicts Scientology as a culmination and grand synthesis of hitherto existing forms of religious belief brought into unity with science itself. As such, it supports the view given by Whitehead of the encompassing comprehension sought by the 'mystic' that she associates with contemporary occultism. In these terms, it is less well understood as the outcome of this-worldly rationalization than as a contemporary form of the charismatic focused on integrating into a single, unified comprehension both this- and other-worldliness. Within such a vision, science becomes integrated into a religious sensibility at least as much as religion is accommodated to science: if religion is scientized, science is also sacralized – and if it is indeed the case that mystery becomes formulae, then it can also be said that formulae become mystery. Such a unified vision is as supportable by reference to Hubbard's 'new language' as much as any practical therapeutic intent, as I will try to show by looking in some detail at his essay, 'Philosophy wins after 2000 years' (Hubbard 1996: 68-73; originally written in 1965 according to the source). Like the creationist pamphlet discussed in Chapter 4, there is so far as I am aware no particular reason to think this essay of any especial significance to Scientologists; it is just one small item in Hubbard's voluminous output. But again, it is its apparent insignificance that is important, that it can have been written as just another part of an ordinary day's work in the ongoing labour of constructing a cosmology for the modern world.

Star Stuff and Clay Feet

In this essay, Hubbard (1996: 69) presents Scientology as the culmination of a process of thought and discovery that goes back to the Ancient Greeks. Their 'natural philosophy' is linked directly to modern science except that their intention of 'understanding … the spirit of man and his relationship to the universe' has been forgotten. They sought to show that 'man was a spirit clothed in flesh', but they lacked the 'higher mathematics and electronics' that has only become available with the advent of modern science. However, this has been misused in the modern era to 'build … atom bombs to wipe out the mankind no one had ever understood'. But now, with the coming of Scientology 'the goals of Greek philosophy live again' and, using the methods and knowledge of modern science, 'after a third of a century of careful research and investigation [Scientology] can answer, with scientific truth' the fundamental questions about our nature 'and prove the answers'. So, as it turns out, he (1996: 70) says, the Greeks 'did not fail' after all:

> They laid … a firm foundation on which to build. And two thousand and more years later we can furnish all the evidence they need. And that evidence and its truths and its great potential of betterment for the individual and all mankind are completed work today in Scientology. We have reached the star they saw. And we know what it is. You'll find its value when you become a Scientologist, a being who has come to know himself, life and the universe and can give a hand to those around him to reach the stars.

In this passage we can see something of the same rhetoric Lessl (1985) identified in Sagan's *Cosmos* (see Chapter 3): a synecdochic reduction employed to create a sense of consubstantiality. This is apparent in a number of aspects of the story Hubbard sets forth including: the narrative construction of history as the unfolding developmental logic of scientific knowing culminating in Scientology; the characterization of 'man' employed as a single, unifying identity for all human beings; and in the imagery of the stars to signify a cosmic and cosmological vision of transcendence. In this way, the 'two directions of transference' Lessl specifies are worked into the narrative. Science – and Scientology by virtue of being equated with it – is humanized through being depicted as the 'grand narrative' of human history and development. Science/ Scientology is therefore within us all – it is '*we*' who have reached the 'star' the Greeks 'saw'; 'we' are all involved not just some select group of people, a unification Hubbard underlines by switching from the third to the second person, 'you'. Thus, the individual reader is invited to include themselves directly in this grand cosmic achievement – at least if they 'become a Scientologist'. Meanwhile, the human is scientized/ Scientologized, because it is this 'philosophical' dimension of human existence that is made the centrepiece of historical development. The point of history, of all human life becomes the achievement of this end, the attainment of a goal that is set for us by 'natural philosophy', that is by science/Scientology. Science/Scientology is the point of existence, just as what it means to be human is to achieve scientific/ Scientological understanding – and the point of it all is to 'reach the stars'.

So, using similar metaphorical and metonymic figures as Sagan, Hubbard likewise offers a grand cosmic vision to capture our imaginations and define our existence and ourselves. The voice of the scientific priest echoes and resonates in Hubbard's own.[4] But there is another tone to Hubbard's voice that works in counterpoint with the metaphor of cosmic journeying, although it also draws on a characterization of science. This is heard in the mention of 'misusing' science to build bombs, but is more fully developed in a passage that follows the imagery of 'stars'. Referring to fundamental questions about life and death he (1996: 70) says,

> Materialistic sciences have sought to invalidate the entire field by shrugging the problem off with the equally impossible answers that one is merely meat and all life arose by spontaneous and accidental combustion from a sea of ammonia. Such "answers" sound more like pre-Buddhist India where the world was said to be carried on seven pillars that stood on seven pillars which stood on a turtle and, in exasperation to the child's question as to what the turtle stood on, "Mud! And it's mud from there on down!"

Here, then, Hubbard constructs a contrast between Scientology and 'materialistic sciences', which in the surrounding context of the essay is also a metaphorical contrast between 'stars' and 'meat'/'mud'. The importance of this is that the imagery of stars and its interconnection with science is given a different spin by being placed within a cosmology in which science itself is presented as having a dual character. Hubbard draws a line of demarcation to exclude 'materialistic science' from true science. It is only the latter that connects us to the stars; the former keeps us firmly stuck to the ground with feet of clay. Thus, 'materialistic science' is part of the profane world; only true science (Scientology) is part of the transcendent order. Mobilizing this demarcation enables Hubbard both to identify Scientology with science and yet keep it apart from representations of science that might seem contrary to Scientology teachings – they can always be defined as 'not really' science or as 'improper' uses of science (see also Wallis 1976: 234-236). Thus, Scientology is made to be both of science and not of science, just as science is made to be part of Hubbard's cosmic vision and yet separable at least in certain respects.

Significantly, the version of science Hubbard equates with Scientology is precisely *not* the version Wilson uses to characterize it: in this essay, 'science' is not a standardized, routinized, instrumental procedure; it is rather a grand cosmic vision that reaches beyond the known world into another realm signified by the stars. The instrumentalized version of science is explicitly rejected. However, this is not to say that Hubbard nowhere adopted a 'this-worldly' model of science – indeed in places he presented himself as what amounts to an ideal-typical empiricist as when he (1996: 52) claimed it was purely the observed evidence that led him

---

4 I am not arguing that Hubbard drew directly on Sagan. As argued in Chapter 3, rhetorics of science may be turned to 'non-scientific' ends using similar rhetorical means. The mid-1960s was the period of the 'space race' and 'space age' imagery was popular (Bell 2006). Hubbard's past life as a science fiction writer was possibly relevant too.

to discover the 'Track', or when he stressed the 'workability' of Dianetics. The point, then, is not to argue that we should replace a disenchanted view of science with an enchanted one, but rather to recognize that *both* versions of science have some purchase in modern culture. We are neither solely disenchanted nor solely enchanted by science; views of its meaning from both sides are available to us as cultural resources and in their elaboration and articulation people, scientists and non-scientists have invented rhetorics that contribute to the repertoires of argumentative techniques for and against either version. Dis/enchantment is not an answer for us but a question, something to think about and to think with, and in the course of thinking (which is also arguing) some of us advocate one side over the other and vice versa – while some of us swing both ways. Sociologically, however, we need to recognize and theorize both and for this we need a re-crafted rationalization that can capture the dynamic interplay. Contemporary religion is only one sphere where such interplay takes place; another is perhaps the logical site to visit after Scientology – science fiction and fantasy.

Chapter 6
To Be Continued: The Magical Power of Super Science

The upshot of the discussion so far is that PMS needs to be understood as informed not by a singular process of disenchantment, but as a mixed discourse involving both disenchanted and enchanted versions that provide contending representations, resources and rhetorics for the construction and articulation of alternative visions of science and its human significance. The last two chapters have focused on contemporary religion as one cultural sphere in which such visions are manifest. Here, established religious forms seek to demonstrate their transcendent capacity by claiming to encompass science, while newer religious forms claim an even higher transcendence that encompasses both traditional belief and modern science in a grander cosmological scheme. However, both exploit the discursive divisions within science that arise from the tensions between its formal and substantive rhetorics, and make use of its enchanted public self-representations that provide resources for speculative extrapolation towards an intelligent designer, whether of an external (God) or internal ('thetans') nature. In either case, they offer their followers an enchanted image of self constructed from syncretizing religious and scientific discursive resources, with the prospect of salvation in a condition that accentuates this-worldly or other-worldly wonderment: stand in awed humility at the spectacle of the miraculous creation in which you yourself participate; or revel in your own miraculous capacity to create a world of your own.

However, whilst these contrasting modes of wonderment are found in the contemporary religious sphere, they have also become manifest in another arena of the social that presumably falls into Weber's 'aesthetic' value sphere: science fiction and fantasy. In this, the paradox of consequences Weber anticipated – that the 'irrational' would become rationalized – has taken place. However, the outcome is not quite what he might have expected, because he did not seem to anticipate the further abstraction of the charismatic into a realm of imaginative speculation in which science provides fuel for transcendent enchanted narratives, even as it is also the focus of critical scrutiny for its perceived limitations. In science fiction and fantasy, the mixed public meanings of science as contrasting potentialities of enchanted and disenchanted scenarios are played out as possible futures (and alternative presents) by extrapolating on the divided promise to bring about technological utopia or dystopic wasteland. Here, Romantic vision and materialist instrumentalism meet in perpetual thematic counterpoint, enabling and encouraging multiple narrative variations from their interplay – not two cultures but one, internally-riven, its two sides co-dependent argumentative siblings, mutually informing and shaping as they wrestle and writhe.

To illustrate and explore this 'rationalization of the irrational', I use the example of American superhero comics. There are many reasons why superheroes are far from representative of science fiction or fantasy in general, but they are nonetheless a phenomenon of modern popular culture that, in the serialized comic book form, owes a fundamental debt to science as a resource of imaginative speculation and verisimilitude (Locke 2005; see also Reynolds 1992) and as such are a part of PMS in need of sociological understanding. Moreover, there are some grounds on which it can be suggested that superheroes provide an example of certain critical features of science fiction/ fantasy and to make the case for this, I begin with a discussion of some more general matters. Two basic points need to be made: first, that a central feature of the representation of science in science fiction/ fantasy is that it is, in the words of John Campbell Jr, 'the magic that works' (quoted by Whitehead 1974: 572); and second that such a representation plays a crucial role in the 'rationalization of the irrational' as it leads to efforts to meld together science and magic. This appears at one level in notions of scientized paranormal powers and at another in the construction of alternative realities bearing some likeness to the 'real' world in which the co-existence of science and magic is given conceptual justification. These features characterize – indeed, arguably, define – superhero comic books. However, out of this genre has come a further, paradoxical developmental consequence: the reversal of the relationship between 'fact' and 'fiction'.

Science as Magic

That science fiction presents a context where science is viewed as magical is outlined by Whitehead (1974: 568) in her discussion of contemporary occultism, which she sees as having a 'natural affinity' with science fiction for two main reasons: the freedom of the 'literary imagination' unleashed in modern culture to, as it were, think the unthinkable; and the charisma of positivism. As regards the first point, the literary imagination that flared into science fiction/ fantasy was concocted in the pulp magazines popular in the United States during the early twentieth century, so-called because they were printed on the cheapest possible paper made from wood pulp (Steranko 1970: 14). Cheaply produced and, at ten cents an issue, cheaply sold, their content specialized in what is often considered the cheapest literary material including a variety of genres, especially crime, horror and adventure that had characterized the similarly cheap and downgraded Penny Dreadfuls popular in Britain during the mid to late nineteenth century (Barker 1989). The pulps sold in high numbers in an intensely competitive market and writers, typically paid a cent a word (Steranko 1970: 27), cranked out stories at a fast rate in a mode of production that facilitated two contrasting outcomes that to an extent still characterize comic book story-telling today. One – the standardized modernist-instrumental approach (Adorno 1991) – tended to rely on proven formulae leading to repetitive plotting with only slight variations on a theme; the other – the inventive postmodernist-playful approach (Baudrillard 1998) – sought to innovate, to find a character, story-line or setting that would define and occupy new symbolic territory in the prevailing economy of signs (Lash and Urry 1994). The pulps, then, provided a cultural milieu that to an extent legitimized stretching the boundaries of popular story-telling, enabling the literary imagination some room for exploration and, although Steranko's (1970: 14) hyperbole – 'every kind of story imaginable, no plot was too remote, no idea too fantastic' – may exaggerate, it is indicative of the pulps' self-presentation.

Among the most fantastically self-styled was that introduced to the pulps in 1926 by Hugo Gernsback, initiating what he considered to be a form of literature sufficiently different as to require a new designation (Turney 1998a: 107). The term he coined, 'scientifiction' (Clute and Nicholls 1993) did not last but the genre of science fiction did and into its fantastical mix, states Whitehead (1974: 568), 'themes and notions taken from the Occult world' easily blended. But this stood alongside what she calls the 'rampant' positivism of the early science fiction community such that:

> The unarticulated charisma with which the Positivist endowed science and technological achievement was but a hair's breadth away from the more expressly magical fantasies which the science fiction writer wove into it. In science fiction creations the Positivist could entertain himself [*sic*] by hovering on the brink of the fantastic while the Occultist could find in the same creations the possibility of bringing the super-sensible realm down onto the plane of hard, commonsense factuality. (Whitehead 1974: 569)

Science fiction then occupies an intermediary ground between 'real', this-worldly science and the other-worldliness of the occult, a space of 'the improbable, the charismatic, and the "unreal".' One indicator of this is the manner in which the science fiction writer 'articulates … the magical' through re-interpreting it in modernist terms, for example by re-crafting magical powers into paranormal abilities and attributing these to 'naturalistic' processes such as mutation or alien influence. So invested with a gloss of scientific credibility or 'verisimilitude' (Whitehead 1974: 570), they become scientized super-powers – still magical, still charismatic, but with an aura of possibility and potentiality through connection to the world of technoscientific wizardry. Such speculative visions, although products of imagination, of the reach of the charismatic into fantastical abstraction, nonetheless retain a quality of believability because, however superficially, they are grounded in the 'hard' reality of the scientifically known and the technologically do-able.

Or at least they do for some, for as Whitehead (1974: 571) recognized other possible stances are opened up by the tensions arising from rationalization, specifically 'the pseudo-scientist' and 'the debunker', the one like 'the occultist' crossing the 'thin epistemological boundary' between this- and other-worldliness, the other a 'staunchily anti-charismatic determinist'. These stances constitute rhetorical alternatives, *personae* (Campbell 1975) generated from argumentation as membership categories that are then available to be deployed in the course of debates about the meaning of science – and they may be used by both or all sides in dispute (Hess 1993, Dean 1998, Denzler 2003). Their divisions are paralleled within science fiction itself where similarly contrasting stances are taken up in debates over the definition of the genre. Advocates of 'hard' science fiction (sf) position themselves against 'soft' sci-fi, while 'science fantasy' is located somewhere yet further afield in a rough correspondence with Whitehead's ideal-typified 'debunkers', 'pseudo-scientists' and 'occultists'. Taken as a whole, these definitional debates show that science fiction occupies a liminal zone (Martin 1981, Ingram-Waters 2009), the fuzzy, uncertain and shifting boundary separating the 'factual' from the 'fictional', the 'real' from the 'unreal'. As such it results from a paradox of consequences: science, in defining itself against its 'other', contributes to its expansion and elaboration. In terms of rationalization, the 'irrational' is not eliminated but emerges into greater prominence, both because it grows in size with every 'factual' claim and because it is fed by the drive toward imaginative speculation that science itself encourages. Disenchantment produces and needs enchantment.

'Fact' Begets 'Fiction'

If, as Shapin and Schaffer (1985) argue, modern science was constructed in its inception as a social space in which to establish 'matters of fact', the corollary was that it should also establish 'matters of non-fact'. So, in seeking to determine by experiment the factual validity of specific claims about the possibility and nature of a vacuum, Boyle was also adding to the stock of things designated 'non-fact' by rejecting some claims about the phenomena 'witnessed' (such as those made by Hobbes), as has every other natural philosopher or scientist in his wake. Similarly, half a century earlier when Francis Bacon sought to establish a general method for distinguishing certain knowledge from the various 'idols' prevailing in everyday life, so did he add to the specification of their range and definition by distinguishing those of the 'tribe', the 'cave', the 'market-place' and the 'theatre' (Larrain 1979: 20, Smith 1998). Thus, not only is the very idea of establishing 'fact' predicated on the presence of a discursive space to which all that is 'not fact' can be assigned, but in attempting to construct systematic means to distinguish with certainty one from the other, the 'non-factual' is both increasingly populated and more subtly distinguished. To pilfer a little from Foucault (Mills 1997), the establishment of a discourse of 'factuality' leads paradoxically to a richer and more variegated discourse of 'fictionality'. Moreover, contrary to what might be assumed, the establishment of something as 'fact' does not necessarily lead to the elimination – the symbolic 'death' (Popper 1972) – of 'falsified' notions, but simply to their re-location within discursive space as they migrate across the territorial boundary

(Gieryn 1999) from 'fact' to 'fiction'. However, as in the example of creationism, they may neither go quietly nor accept they should go at all; hence they may remain matters of argument at least for some. Added to which the traffic across the boundary is by no means necessarily only one way.

Science contributes to fiction in at least two major ways: first in accounts of error; second in inducing imaginative speculation about the nature of 'matters of fact'. Accounts of error were discussed in Chapter 3, where I argued that a link could be made with Weber's distinction between 'formal irrationality', 'substantive rationality' and 'substantive irrationality', roughly corresponding respectively to methodological errors, social interests, and personal factors as dimensions of the rhetoric of science. To maintain the assumption of a single, common, external reality in the face of contrasting versions of that reality, scientists have generated an internal discourse consisting of a variety of standardized verbal formulations for discounting alternative views to their own. Given that scientists in controversy associate their own views with the 'facts', these rhetorics of error provide means of justifying the designation of alternative views as 'non-factual'. So, 'non-factual'/ 'fictional' discursive space increases with every scientific controversy, to say nothing of how it is swollen by the rejection of beliefs that are simply deemed outside of science. There is a certain kind of self-propagation to this as accounts of error stimulate accounts of these accounts within the idiom of science itself, leading to various psychologies and sociologies of error (Bloor 1976). These may be taken up by 'non-scientists' and used to their own ends – for example, as creationists have taken up sociological notions of professional socialization in accounting for what they see as the 'errors' of evolutionists (Locke 1999a; see also Richards and Ashmore 1996). This is one way in which traffic across the divide between (purported) 'fact' and (supposed) 'fiction' may move in both directions, leading to the need for perpetual boundary maintenance, which has helped to generate social space for specialists in debunking like the Committee for Skeptical Inquiry (CSI). In Chapter 7, I look at this trafficking further in relation to Fortean discourse.

The debunkers' task is made the more troublesome by a second way in which the boundary may be breached, through the inducement to imaginative speculation that science encourages both to sustain the 'experimental way of life' (Shapin and Schaffer 1985) by generating new research foci and to provide prospective explanatory accounts of the phenomena so 'witnessed'. Thus, scientists come up with possible realities, some of which – perhaps over the course of time *all* of which (a view taken by Weber 1948b) – will come to be designated 'not factual'. They may, however, in the meantime be taken up and used by the 'literary imagination' for further speculation of an overtly fictional nature. Thus science 'fact' generates science 'fiction'.

However, this begs the question of where the stimulus to the 'literary imagination' originally came from. Here, Campbell's (1987) study of the 'Romantic ethic' is helpful as it opens up an approach to this in terms of rationalization. Campbell's main concern is to fill the absent complementary counterpart to Weber's Protestant ethic thesis concerning the modern drive to consumption or 'modern autonomous imaginative hedonism'. He argues that, while Weber's thesis provides an account of the drive to produce in modern capitalism, it says nothing about the drive to consume. But with the additional production inspired by the Puritan will to work, there must have been something providing an equivalent will to consume or the dynamic cycle of capitalist accumulation would never have kicked into gear. Campbell's proposed solution, as initially unexpected as Weber's own location of the drive to material productivity in the life of the spirit, is that it was the very asceticism of the Puritan ethic that instigated consumerist hedonism. The link in the chain is the imagination and how this was fed by the sober self-discipline of the Puritan way of life. Campbell argues that Puritanism encouraged two countervailing tendencies: suppression of the passions to maintain an outward image of disciplined sobriety; and an intense focus on the prospects of the hereafter, especially the possibility of eternal damnation. The latter stimulated emotional intensity – Campbell points out that Puritans were renowned for their deep melancholies – but the former suppressed any outward expression; the result was a powerful inward turning that stimulated the life of the imagination. In attempting to trace the development of this energized individual imagination from its Puritan source to the condition of perpetual wanting he sees in the spirit of modern consumerism, Campbell takes in the rise of modern fiction so providing a way of understanding the liberation of the 'literary imagination' in Weberian terms. Thus, however adequate for understanding contemporary consumerism (see Gabriel and Lang 1995), Campbell does provide a way of linking the development of Romantic fiction in particular to the process of rationalization. As seen in Chapter 3, Romanticism concerned Weber as something of a counter tendency to disenchantment, against which he advocated a disenchanted outlook as appropriate for the scientist. For Campbell, however, Romanticism is also an outcome of rationalization, which provides a basis for viewing it as an articulation of the charismatic in the aesthetic sphere. As such, the rejection of the disenchanted, formal rhetoric of science found in the Romantic ethic can be seen as stemming from a search for transcendence similar to that articulated in 'occultism' and 'reflexive spirituality' – in Bell's (1976) terms these are all expressions of the modernist 'will to beyond'.

This can be seen especially in science fiction and fantasy. Science fiction has been traced to a variety of historical sources sometimes as far back as classical mythology (Lambourne, Shallis and Shortland 1990), but more pertinent is the view that links it to Gothic literature, notably Mary Shelley's *Frankenstein, Or the Modern Prometheus* (1994; see Aldiss 1986). This is not to deny the significance of earlier sources for both narrative themes and character types – and in the case of superhero comics perhaps the basic story formula (Eco 1979) – but such sources have been taken up and used as part of the broader discourse that science fiction articulates.[1] As its sub-title attests, *Frankenstein* cues a general theme from

1 It might also be suggested that the link to classical mythology is a legitimizing strategy for what was long regarded a culturally downgraded genre, a point also made about superheroes (Reynolds 1992: 53).

classical mythology, but this is significantly re-crafted by Shelley to suit a modern technoscientific context. The general theme, succinctly put by Aldiss (1986: 26) is 'hubris clobbered by nemesis'; its modern application is to insist that there are moral limits to human knowledge crossed at peril, or as stated in the 1936 film *The Invisible Ray*, 'there are some things man is not meant to know' (cited in Woodlief 1981) – and a clearer rejoinder to the disenchanted outlook Weber associates with science could not be given. However, this recurring lesson in science fiction tells only one side of the Romantic ethic; its counterpart is the side stressed by Whitehead imbued with the spirit of positivism, reaching for transcendence through the rocket-fuelled imaginative speculation opened up by science and technology. On this side, far from their being limits to human knowledge, the potential of scientized humanity is limitless – but as much as this might seem to express a disenchanted outlook, it is also born of a sense of wonder:[2] it may indeed be a spirit that seeks to penetrate the mysteries, but only in order to boldly go, like the crew of *Star Trek*, into further mystery beyond.

Superhero Continuity and/as the Rationalization of the Irrational

In science fiction and fantasy, then, the tensions surrounding and informing modern science are articulated and elaborated. Superhero comics provide some illustration of this and also how the tensions unravel towards the 'rationalization of the irrational'. This raises the question of definition – or rather it does not, because even if it could be resolved (Vieth 2001), it is not relevant for this argument whether superheroes are considered science fiction or science fantasy. What counts is that they are part of the fictional/ fantasy space in which modern culture explores the meaning of science. Representations of science are fundamental to superhero comics (Reynolds 1992, Locke 2005; see also Oehlert 1995, Rhodes 2000, Possamai 2006) and as such they will serve for the purpose of examining PMS.

Is it then possible to define the superhero genre? A recent extended attempt to do so (Coogan 2006) is notable in my view for its evident failure – which is not so much a criticism as an expression of sympathy – and one of the ways in which it fails raises a key matter of concern: *continuity*. Coogan (2006: 39) argues that a genre can be defined by its central conventions and identifies as the 'core' of the superhero genre 'three elements – mission, powers, and identity'. Thus, superheroes tend to be motivated by a purpose, such as Spider-Man's dictum that 'with great power there must also come – great responsibility' (Lee and Ditko

2 I do not intend this as a putative definition of sf and nor do I intend it ironically; but it is significant that it has been used as a definition and also been a focus of criticism (Clute and Nicholls 1993), which illustrates my point – both enchanted and disenchanted outlooks are available and articulated as means and focus of argumentation.

1962: 11);[3] they have paranormal powers often connected to their mission; and they have a dual identity living as an ordinary person when not in costume which then functions as a signifier of their 'super' status (Reynolds 1992).

However, Coogan (2006: 46) also mentions other conventions that he does not consider 'core', amongst which is continuity. So, in discussing some characters, such as the planet-hopping hero Adam Strange, who he says 'clearly falls within the science fiction genre' and John Constantine, described as 'a straightforward trench-coat-wearing horror investigator', Coogan (2006: 51-52) states: 'Like ... Adam Strange, John Constantine's placement in a superhero universe might mistakenly be seen as marking him as a superhero. But to do so would be to overvalue the continuity convention and to undervalue the mission, powers, and identity conventions and also to ignore the importance of genre distinction.' This, however, is circular as which conventions are 'valued' is a matter of analytical choice. Coogan chooses to emphasize 'mission, powers, and identity', but there is nothing to stop me emphasizing continuity. To say that this 'ignores the importance of genre distinction' puts the cart before the horse since the existence of such a distinction needs to be established first. As Coogan clearly recognizes, superhero comics incorporate a rich variety of characters from a wide range of fictional backgrounds and the issue of generic distinctiveness is at best subsidiary since supposed genre boundaries evidently present no obstacle. Rather, the co-presence of these characters in the same fictional space or 'reality' is accepted, although it can raise the question of how this is possible – and this is a matter of continuity.

Thus, regardless of whether they can be said to constitute a genre, superheroes occupy a fictional space that publishers, creators (writers, artists and editors) and at least some readers have sought in an increasingly systematic way to define through the construction of what is conventionally called a 'shared universe'. In many respects, this is what continuity is; moreover, the activity of its construction is exactly the 'rationalization of the irrational', as can be brought out through some discussion of the best extended treatment, that of Reynolds (1992: 38-47). Reynolds argues that the term 'continuity' covers three forms of intertextuality that are often conflated: serial, hierarchical, and structural. However, while these distinctions are useful, he does not consider that their conflation arises because they are interconnected stages in a process of rationalization. Rather, he views them structurally which leads to problems especially with his treatment of structural continuity. To resolve these, I will argue that continuity is a phenomenon of mundane reasoning that involves the use of the documentary method of interpretation (Garfinkel 1967) and it is this that results in – and practically

3 There is no established convention for comics citations. As with many popular cultural texts, authorial attribution has not always been paramount, further complicated in the case of comics by multiple authorship often involving writer and artist(s), including: penciller, inker, letterer and colourist – to say nothing of the role of editors (Brooker 2000: 266-279, Varnum and Gibbons 2001). To minimize unwieldiness, I have usually cited by writer and main artist (usually penciller).

constitutes – the rationalization of the irrational. To begin to elaborate this, some discussion is needed first of serial continuity.

Serial continuity refers to the kind of narrative features found in genres like soap opera that define a level of consistency to characters and story-lines across separate appearances – the 'back-story', as Reynolds (1992: 38) puts it. The development of such serialization might be viewed as the crucial feature of popular fictions, indicative of standardized and regularized production formats as emphasized in critiques of mass culture that draw on the standard view of rationalization (discussed in Chapter 1). Seen this way, serial continuity is viewed for the apparent stability or 'conservatism' (Dittmer 2007; see also Hatfield 1998) it gives to fictional forms, which in the case of superheroes especially has been held to stultify creativity to the point of restricting the development of comics as a whole in the United States (Bongco 2000, Locke 2009b). It is then often seen as a controlling concern favouring the interests of producers – that is, comic book companies – rather than creators – writers and artists. The history of superhero comic books could be taken to lend some support to this, since the monthly serial form became more firmly established after the success of Superman, who first appeared in June 1938 in *Action Comics* 1 (Siegel and Shuster 1997[1938]), usually seen as the first superhero comic. This led to the rapid growth of the comic book industry prior to the United States entering the Second World War and *Action* has followed a monthly schedule of publication, continuing more or less uninterrupted to the present day with the *Superman* strip still the lead feature. Further, although the industry as a whole has experienced changing fortunes, the two surviving publishers that have continued to produce superheroes on a regular basis throughout are DC, owners of Superman, and Marvel, who first entered the market in 1939, and it is notable that both have paid increasing attention to the continuity of their respective universes over recent decades.

However, to maintain serial continuity over such a period of time strikes me as quite remarkable and to be fully understood requires something more than simply serving to keep the wheels of capitalist production turning. Indeed, it is often argued to be actively counter-productive in this respect, because it requires too much investment from readers to build up and maintain the level of so-called 'cultural capital' (Brown 1997, 2001) required to make sense of stories with complex inter-woven plots strung out over several issues involving months if not years of back-story. Hence, some argue the market for comic books has been in slow decline directly related to the growth of fan interest in continuity to which publishers have catered for the sake of a captive market, but at the loss of a supply of new, younger readers to replace the existing aging audience (Pustz 1999, Hatfield 2005). If, then, continuity is indeed a 'conservative' process designed to maintain the economic interests of publishers, it would seem rather self-defeating and they ought to have abandoned it, but just the opposite is the case. Understanding this demands something more than the one-dimensional conception of a dominating instrumental logic of production afforded by the standard view of rationalization. At the very least, as Bongco (2000) points out, the interest of readers has to be maintained and it is simply inadequate to assume all of them have been lulled into a state of mental passivity for which essentially the same formula(e) endlessly repeated is sufficient. This is not to say there are no such formulae, but it is to raise questions about what they might be doing, to question whether formulaic repetition equates to reader passivity, dull-wittedness or uncriticality, and to propose that 'sameness' and 'difference' are in significant respects in the eye of the beholder (see Jenkins 1992: 66 on 'double-viewing').[4] Regular readers may see variations that occasional readers do not and this may induce puzzle-solving in relation to what they take to be established continuity, which may then generate further variations and more puzzles.[5] To perceive this as 'conservative' is at the very least partial, since it misses the active, creative work involved in such sense-making.

Superhero continuity is an extraordinary example of this sense-making work involving several generations of creators and readers; it is nothing less than the active work of making culture – 'immortal ordinary society' (Garfinkel 1991) in action – and it employs the documentary method of interpretation. In Garfinkel's (1967: 76-103) usage, this refers to a feature of ordinary, commonsense reasoning in which particular phenomena and events are treated as 'documents' – signs or indices – of an underlying pattern and interpreted accordingly. However, new events may lead to revisions in the assumed pattern, at which point the revised pattern is taken to have been the actual pattern all along. Thus, there is a continuous hermeneutic toing-and-froing in which events are interpreted through the pattern and the pattern is interpreted through the events, but in either case a sense of order is maintained, in which events and actions are treated as being and as having always been in accord with that order – and thereby as rational, accountable, credible and creditable. In a word, the documentary method of interpretation is a procedure for accomplishing rationalization, but this is now to be understood as a matter of ongoing practical sociological reasoning. This can be seen in the shift in superhero comics from serial to structural continuity.

---

4 The view that superheroes are formulaic is common, but there is little obvious agreement as to what the formula is – see Eco 1979, Blythe and Sweet 1983, Reynolds 1992, Bongco 2000, Brown 2001, Coogan 2006; also Lawrence and Jewett 2002. The most sophisticated discussion of the notion of 'formula' with respect to comics is Barker (1989; also 1984, 1993, 1997). Unfortunately, he does not discuss superheroes although his general point that formulae provide readers with resources to think about dilemmas in their lives seems to me broadly relevant. In the case of superheroes, as I have argued elsewhere (Locke 2005), amongst the dilemmas they present are ones over the meaning of science.

5 I do not intend the term 'regular reader' in Barker's (1989) restricted sense. He distinguishes three categories of reader: 'committed', 'regular' and 'casual', and deliberately avoids the term 'fan' which he (1993: 179) argues is an 'industry' category that 'committed' readers of comics reject. Brown (2001: 67), however, claims 'fan' is a readers' category and some who might qualify as 'committed' in Barker's sense adopt it as a non-pejorative self-description. This is not a matter to try to resolve here, so suffice it to say I use both terms as convenient.

Metatext(s) of Science and Magic

Structural continuity refers to 'the entire contents of the DC or Marvel universes' (Reynolds 1992: 41). However, it is more than just the events as recorded in published stories as it 'also embraces those elements of the real world which are contained within the fictional universe', including 'actions which are not recorded in any specific text, but inescapably implied by continuity' – such as in Reynolds' example, 'what was the name of Superman's grandfather?' Although this character may never have appeared or been mentioned, continuity seems to imply he existed and therefore potentially the information could be presented or made the basis for new stories. For Reynolds (1992: 43), it is structural continuity that is critical to reader involvement through what he calls an 'ideal ... metatext: a summation of all existing texts plus all the gaps which those texts have left unspecified'; that is, therefore, a single underlying pattern. As he points out, however, the metatext is necessarily an abstraction, as it is unlikely that anybody has read all the possible texts to be included and new texts are being continuously added by the monthly publishing schedule – new events then to be interpreted in terms of the pattern or through which the pattern may be revised. And here we hit a snag in Reynolds' account.

For Reynolds (1992: 45), 'metatextual structural continuity' defines the nature of superhero comics as 'myth', a term he uses in a Levi-Straussian sense as 'machines for the suppression of time' (see also Eco 1979), such that 'continuity is a *langue* in which each particular story is an utterance'. But this begs the question of why it should be assumed that there is only one metatext. Reynolds (1992: 43) is explicit that it is an abstracted idealization that is predicated on an 'ideal fan', but given that by his own account such a fan does not and cannot exist, on what grounds can this idealization be justified? The metatext can in effect be defined as whatever the analyst chooses, because there is no ultimate arbiter to which to resort. So, for example, speaking as a fan (if far from ideal!), I can easily dispute Reynolds' claim that Superman's grandfather is implied by continuity and do so *in ways that are in keeping with continuity*. There are numerous possibilities consistent with the logic of DC's universe and they run the gamut from magic to science: maybe Superman's Kryptonian parents were created *ex nihilo* by the magical fifth-dimensional being Mr. Mxyzptlk as part of some grand temporal trick on the Man of Steel; maybe the Time Trapper, a super-mad-scientist-cum-mage who lives at the end of time constructed an alternative time-line in which Superman's parents effectively started life fully matured; maybe they were concocted in a test-tube from artificially engineered DNA in a robotized Kryptonian factory; or maybe the whole 'planet Krypton' idea is an implanted memory, just an imaginary story.

So does this mean there is no metatext? No, it means there are *multiple versions* of the metatext (Hatfield 1998). Reynolds is right to liken continuity to a language, but his structuralist view of language misleads him. Continuity has the same odd feature of language in being a bounded infinity in which any new utterance is possible even though for the most part the same old utterances are repeated. Like a Foucauldian discourse, continuity is the means of generating utterances, but it enables as much as it constrains. New things can be said because discourses are informed by tensions that generate puzzles. Tensions enable different versions to be imagined and thought through using the documentary method of interpretation, as I have just tried to demonstrate in my made up example.

Of course, it might be objected that this *is* a made up example and so has nothing to do with what comics readers and creators actually do. Fortunately, however, Reynolds (1992: 45-46) provides a case of ostensibly real[6] readers doing just this kind of thing in letters of comment (LOCs) concerning the powers of the Human Torch, a member of the superhero group, the Fantastic Four, whose body bursts into controlled flame. In one such LOC, a reader refers to an earlier letter setting out their 'theory on the Torch's flame: namely, coating his fireproof body and clothing with "photo-chemical endo-thermal reaction utilizing quasi-volatile gasio-solids and mentally dominated oxygen-rich plasmas".' They then state that a 'brief flurry of controversy stirred Marveldom Assembled' (a reference to Marvel readers) and from this '[i]t has been suggested that the Torch and the X-Man Iceman become their respective elements, fire and ice', a view criticized on the grounds that 'such creatures' would cease to live if their powers were removed and this does not actually happen in the comics. The reader then proposes 'teaming up' the two heroes suggesting a possible plot-line.

As Reynolds rightly states, this kind of involvement by readers is far from unusual and is indicative of the success of the strategy adopted in the 1960s by Marvel's editor, Stan Lee to encourage such participation (Pustz 1999, Raphael and Spurgeon 2003). The example illustrates two kinds of involvement in continuity puzzles: one at a more 'serial' level concerned to connect existing disparate events as the basis of a new story; and the other at a more 'structural' level to propose a purportedly scientific explanation for the Torch's powers. This explanation may well be 'bogus', as Reynolds (1992: 46) assures us, but its importance resides in what it shows about how the superhero universe is viewed cosmologically. It is notable both that some other fan(s) had evidently presented an alternative possibility and also that this is not challenged as to its cosmological legitimacy, but only in terms of how it fits with established continuity. Reynolds describes the alternative explanation as 'more openly magical' although it needs to be acknowledged this term is not used by the letter-writer. Nonetheless, that magical explanations are accepted in superhero universes is apparent from such features as the presence of characters whose powers are of an explicitly 'magical' nature, like Marvel's erstwhile 'sorcerer supreme', Doctor Strange. Thus, science and magic exist side by side and are accepted as having equal legitimacy as resources out of which alternative interpretations and arguments may be generated.

So, different versions of the metatext are possible and one thing that enables this is the accepted availability of both science and magic as legitimate resources

6 No slight is intended by this but it needs to be recognized that, as comics readers themselves sometimes suggest, producers of comics might have made up LOCs.

of understanding. Reynolds (1992: 47) himself effectively acknowledges the first point when he states:

> The Torch has baffling and inexplicable powers, but as these powers are real within the given structure of continuity, the job is not to make those powers less baffling, but to explain them within the parameters which continuity allows. *Yet continuity is also something malleable, and constantly in the process of being shaped* by the collective forces of artists, writers, editors, and even the critical voices of fans. (Emphasis added.)

This is precisely the point: continuity is not fixed, not even as an abstracted idealization. There is no *langue*; there are only utterances (Vološinov 1973, Billig 1997). Each utterance is a new event to be fitted to the pattern, but the pattern itself is revised by each new event – and moreover, this is true for *each* reader and creator. Routinely, however, superhero texts are *assumed* to constitute a continuous reality, which is further assumed to be the same as 'real' reality is assumed to be: having an underlying pattern that is constant, singular and coherent (one idealized metatext). But there is no guarantee that everybody understands that reality in the same way any more than there is any guarantee we all understand 'real' reality the same way. We tend to assume 'real' reality is shared and the same for all, and comics readers and creators tend to assume there is only one continuity, but just as we are continuously confronted by 'reality disjunctures' (Pollner 1975, 1987) in the 'real' world, so also are comics creators and readers confronted by 'fantasy-reality disjunctures' in the form of different versions of continuity. Hence, they argue about it just like in 'real' life.

Here, Reynolds' observation that fans' voices are 'critical' is important, as one focus of this concerns science and magic. Continuity is not a unity but a multiplicity containing both scientific and magical points of reference and potentials, which raises the issue of their relative meanings and roles within superhero universes. Reynolds argues they are treated as effectively indistinguishable, uniting to form a sense of wonder, but it is clear that this is not the whole story, because they may also be treated as separate explanatory resources. Although continuity admits both as legitimate means of explanation, they are not necessarily united or seen as identical – and their status as 'bogus' or otherwise is quite beside the point (see also Chapter 1) unless it is made the point as part of an argument about continuity. For example, in Superman's first appearance, a compressed evolutionary explanation for his powers was presented, stating that he is from a species of being 'millions of years advanced' of humans in its 'physical structure' (Siegel and Shuster 1997[1938]). Originating from the planet Krypton, Superman is ostensibly an alien although the evolutionary characterization suggests he is an advanced human. However, this explanation disappeared in later comics and, when DC reconstructed the continuity of their universe in the 1980s, it was established that Superman's powers come from absorbing solar energy from earth's yellow sun (Byrne 1986). Whether or not the reason for this change is because the 'evolutionary' explanation is now considered 'bogus' is not clear, although it is the case that the superman heroic type was criticized in the post Second World War period for its apparent resonance with the Nazi *Übermensch* (for example, Wertham 1955). This type of authoritarian reading is now regularly used in the comics themselves to present critical takes on the superhero, such as in state-funded 'super soldiers' where genetic engineering substitutes for Superman's spatially-displaced temporal evolution (for example, Veitch 1985-1986).

This is one way in which the Romantic questioning of the meaning of science informs superhero comics, as is the ubiquitous figure of the mad doctor (Skal 1998) as a type of super-villain. In such characters, it is often the case that science and magic are combined, such as in Marvel's Faustian villain Doctor Doom, who is both super-scientist and mage. Equally, however, Doom's heroic counterpart, Reed Richards, leader of the Fantastic Four, although something of an archetypal 'saviour' scientist (Haynes 1994) also combines science and magic but in a less explicit fashion. For Richards, magic is troublesome, something he cannot readily accept or contemplate (for example, Waid, Wieringo and Kesel 2003) and part of Doom's 'evil' status is that he breaches what is for Richards the normative boundary separating science and magic, as seen in Doom's origin in *Fantastic Four Annual* 2 (Lee and Kirby 1964: 9-10). As college students, Richards chanced upon Doom undertaking 'forbidden experiments … trying to contact the nether world' using technoscientific means involving 'matter transmutation and dimension warps' that Richards immediately understood sufficiently to point out some errors in Doom's 'equations'. The arrogant Doom proceeded regardless and the resulting accident led him to don the armoured mask that signifies his villainous condition, thereafter seeking twisted vengeance on Richards. Doom's 'evil', then, arises from having breached in an explicitly magical direction the conventionally accepted limits of knowledge that Richards represents. However, Richards also owes his paranormal powers to a breach of the limits of established knowledge, when he and the rest of the Fantastic Four were bombarded by unknown 'cosmic rays' during an unsanctioned flight in an experimental rocket-ship. Thus, Richards's own 'super' status came from a source beyond the limits of current knowledge and, like Victor Frankenstein, he bears a moral burden in consequence, as one of the Four, Ben Grimm was turned by the rays into the tragic, monstrous Thing. So, despite their scientized verisimilitude, the cosmic rays remained mysterious and enchanted. In both Richards and Doom, then, enchanted and disenchanted features are combined to enrich the characters and open up narrative possibilities that articulate and further elaborate the dilemmas in the unresolved arguments about science and its relation to magic.

Mad scientists also represent a form of critical opposition to modernist science (Toumey 1992). For example, Superman's arch-nemesis, Lex Luthor stands opposed to the modernist utopian vision of science that the alien man-god represents, but just as Superman can be described as a being of both magical science and scientized magic, so too can his mad scientist counterpart. Luthor is an ordinary man with no super-powers, but his 'super' understanding of science enables him to develop, albeit in vaguely specified ways, technologically enhanced means to challenge Superman's god-like powers. Thus, Luthor represents the 'magic that works' – a magical technology to oppose the scientized magic of Superman – but in using scientific means to the 'evil' end of defeating the symbol of hope for a scientized utopia, he also represents its critical 'other'. Moreover, whilst Luthor's villainous status might be taken to imply that the critique is both cognitively and normatively wrong, not only is it never fully overcome – the proverbial battle is 'never-ending' – but the presence of such characters opens up the potential to develop the critique further. This can be seen in the rise of 'anti-heroes', such as the mutant X-Men. Like Superman, the X-Men owe their paranormal abilities to a speculative vision of the potential of science to produce enhanced humanity, although in them the evolution is envisaged not through time displaced into space, but through technological means in the form of exposure to nuclear radiation. Unlike Superman, however, the mutants stand in doubtful, fraught relation to wider society to whom they appear less as god-like saviours and more as monstrous freaks, by-products of an uncontrolled laboratory experiment with uncertain and potentially threatening consequences. The X-Men first appeared in 1963 (Lee and Kirby 1963b), some 30 years after Superman was first dreamed up by Jerry Siegel (Catron 1996), so perhaps the contrast between them is indicative of a shift in PMS in the post-atomic period towards the growing anxieties of 'risk society' or even the preternatural rumblings of the postmodern condition. But even if so, they resoundingly echo the earlier Romantic sentiments regarding the problematic moral status of advancing knowledge.

Moreover, to look to wider social structural processes is to ignore the internal workings of continuity. The combining of science and magic in individual characters and their conflicts is indicative of the tensions surrounding science in modern culture regarding whether it enchants or disenchants and if it is a good or bad thing in either case. However, while these tensions provide rich resources for fictional constructs and scenarios to be imagined, they are also treated as potentially conflicting alternative means of persuasion and resources with which to construct arguments, as seen regarding the Human Torch's powers. Starting in the 1960s, this led to a growing effort to rationalize superhero universes in the specific sense of elaborating a cosmological order in which both science and magic were treated as co-existing. Once characters whose powers were attributed to different kinds of sources were presented as occupying a shared reality with common biography in a single emerging history, then the continuity puzzle of how their co-existence is possible was prompted. Importantly, however much such a question might only be posed by virtue of the fantasy world in some fashion being a version of our own 'real' world in which the discourse of disenchantment that opposes science to magic has some cultural purchase, the response has not been to deny the veracity of non-scientific sources. Rather, it has been to treat the disenchanted viewpoint as an argument to be argued with and about and over. As was seen in the case of the Human Torch, both scientific and magical versions were treated as legitimate possibilities within the rules of continuity enabling readers to construct different versions of the metatext. This led to an argument over which made more sense in relation to established 'facts' of serial continuity – that is, which underlying pattern better fit the observed events – but it did not lead to an argument that magical explanations have no place at all. The rules of continuity allow both possibilities, but given this the next level of puzzle is to ask how the two fit together as part of an underlying pattern that fits established events even as those events may then be reinterpreted through this pattern (or 'retconned', short for 'retrospective continuity').

From Realistic Magic to Magical Realism

Thus, the pursuit of structural continuity has led to efforts to establish some form of transcendent coherence on a grander scale and it is here that the 'rationalization of the irrational' most fully emerges. There has been to this broadly a three-part development: first, the emergence of 'cosmic' level characters, powers and storylines taken to define a basic set of constituent elements of a given superhero universe; second, the attempt to systematize a given universe by establishing an order between the cosmic powers and their relationship to the 'ordinary' superheroes; third, the treatment of a superhero universe as reality itself, thereby reversing the move from science 'fact' to science 'fiction' and doing so on the grounds that the 'magical' constitution of the fantasy world is the underpinning of reality, that in other words reality *is* magic explicitly understood as symbolization. This is exactly what Whitehead anticipated as the next stage in the abstraction of the charismatic, but it can be more fully understood as an outcome of the workings of mundane reasoning constructing coherence through the documentary method of interpretation in the context of a culture that incorporates both the scientific and the magical as available resources and means of argumentation.

'Cosmic' characters emerged out of 'hierarchical' continuity. In general, this refers to the relative positioning of superheroes in terms of their powers and moral character – crudely, who would win in a fight. In this respect, hierarchical continuity is an outcome of rationalization applied to serial continuity once characters began to cross-over and interact. In part, this follows from already established events as Reynolds (1992: 40) points out: if a hero has beaten a villain who then appears to beat another hero, then it follows that the first hero is more powerful than the second. However, source and type of powers have a bearing on this since some are more clearly matched than others. For example, the Marvel characters, the Hulk and the Thing are both super-strong, so their relative positioning is more straightforward than a character like Doctor Strange who has no super-strength but has all the other-worldly power of magic at his disposal. So, this raises the continuity puzzle of where magic fits in relation to ostensibly non-magical powers, which provides an impulse towards a more encompassing level that incorporates both.

This is also stimulated by another aspect of hierarchical continuity that Reynolds (1992: 41) calls the 'extra effort, the moral determination to go on fighting even when apparently beaten.' Even in the earliest Superman stories this was often the crucial turning point, when the apparently beaten hero would manage to overcome a final obstacle through some extra effort. But this contributed to a progressive, if somewhat haphazard inflation to Superman's strength and abilities, so that while he began faster than a speeding bullet and able to leap tall buildings, he ended up faster than light and able to fly. This prompted continuity puzzles as to the extent of his abilities and questions about their 'scientific' possibility. Thus, although hierarchical continuity works to establish a pecking order of powers, it also pushes heroes beyond their established limits, which can challenge the pecking order, prompting continuity revisions and lead to spiralling inflation of power-levels inviting grander challenges against ever more powerful threats. Thus, 'cosmic' level characters were eventually introduced with powers that in both origin and capacity far surpassed the 'ordinary' heroic level and made up more fundamental constituent parts of the fantasy universe as a whole. In effect, there is an equation between 'power' and 'reality constitution': to have more power means ultimately to be able to affect and use deeper levels of the constituent features of reality. Power then is transcendence; it is charismatic enchantment.

Thus, the development of the cosmic level can be understood as a rationalization of the irrational in the sense of a progressive abstraction towards a more universal level, encompassing all established serial continuity and prompted in part by hierarchical continuity itself arising from the rationalization of serial continuity. This reaching toward an universal level threw up further continuity puzzles beyond merely the systematization of individual character's biographies and the pecking order, as it raised questions about the fundamental make up of the universal order, especially concerning the co-existence of science and magic. Reynolds (1992: 16) mentions one example involving the Marvel heroes, Thor and Iron Man. Thor is Marvel's version of the Norse god of thunder, while Iron Man is a modern technoscientific knight in armour. Their co-existence was established at an early stage when both joined the superhero team, the Avengers, whose comic became one of the most central in the development of cosmic level story-lines and characters, such as the intergalactic war between two ancient, technoscientifically advanced alien races, the Kree and the Skrull in which Earth became the battleground, serialized between *Avengers* 89 (Thomas and Buscema 1971) and 97 (Thomas and Buscema 1972). Such events helped establish a history of the Marvel universe within which individual character's biographies were located. However, the presence of Thor, an essentially mythological – and from a modern scientific point of view, fantasy – character within this cosmic space-opera might seem anomalous. Added to which, in Thor's own serial continuity, the existence of a pantheon of Norse gods occupying the other-worldly realm, Asgard had already been established, together with other pantheons and dimensions of existence. So, when Iron Man was sometimes depicted puzzling over how he as a good sceptical scientist could accept the existence of this magical being, he was giving voice to a kind of continuity puzzle prompted by the existence of magic in a supposedly disenchanted world. But just as in the case of the Human Torch, the adopted resolution was neither to abandon magic nor dissolve the distinction with science, but to advance a more transcendent order into which both could be situated.

This is apparent from the approach made explicit by Marvel editors, when they set about attempting to formalize the continuity of the Marvel universe in the 1980s. In a preface describing their procedure, entitled 'Scientific Method', it was stated that:

> The premise around here is that the laws of physics apply to superhuman powers ... unless the powers are derived from certain unexplainable phenomena (magic, psionics, extradimensional energies) that must somehow coexist with the science we know. Consequently, unless someone's powers stem from "mysterious" sources ... they are subject to the same laws of physics we are. (Gruenwald 1986: inside front cover)

The existence of 'unexplainable' and 'mysterious' sources is accepted and magic given a place rather than treated as either erroneous or to be explicated in scientific terms. It is incorporated into a more transcendent 'reality', so that, for example, other-worldly lands such as Asgard are defined as 'extradimensional' components of a 'multiverse' that effectively allows space for any kind of imagined possibility. Thus, although there is scientization, it is limited. This is in keeping with the use of the documentary method of interpretation, which does not necessitate abandoning sources of apparent incoherence, but may strive instead to account for them in a manner that satisfies the sense of rational order by virtue of constituting that order as such through this accounting. The 'mysterious' is accepted as a part of the fantasy world alongside the 'laws of physics' through so constituting it as 'mysterious'. But there is nothing 'irrational' about this because this is what the order of the reality is made to be.

Accordingly, magic is not destroyed in modernity or argued out of existence by the formal discourse of disenchanted science; it is instead relocated into a discursive space where enchanted potentialities remain as speculative possibilities in the imaginative creation of fictional worlds. But there is no necessity why it should stay there and the further rationalization of the irrational has been to bring the ostensibly fictional back into the discursive space of the factual, such that reality itself becomes understood through the magic of the superhero universe – a paradox of consequences illustrated by the example of Alan Moore's *Promethea*, a series that ran for 32 issues from August 1999 to April 2005. Moore wrote the stories and a number of artists worked on it including, J.H. Williams III, Mick Gray, Jose Villarubia, Jeremy Cox and Todd Klein. Their work was vital to the story-telling, but *Promethea* was used by Moore as a vehicle to expound his personal cosmological vision and for convenience I refer to his name only in what follows.

By the 1990s, structural continuity had become the norm for superheroes so that any new publishers entering the market tended to do so with the idea that their characters already existed inside a complete universe. Producers tended to think not so much in terms of individual heroes and series, but of intersecting characters and story-lines. Thus, *Promethea* was situated as one series published by America's Best Comics (ABC) and, like other titles in the ABC line, was ostensibly a superhero comic. However, in a notable terminological shift, Moore coined the term 'science heroes', which he applied to the character Promethea even though she is explicitly a being of magic. Thus, the effective integration of science and magic was already signified. However, in a more radical departure, Moore used *Promethea*, ostensibly a work of fiction, to describe the nature of 'factual' reality, outlining a conception of 'fact' as another type of 'fiction' – and both as magic.

There is in this a similar attempt to articulate a transcendent understanding as seen in the discourse of 'reflexive spirituality' discussed in Chapter 5 (see also Locke 2008). Here, I will limit discussion to Moore's use of synecdoche, as this shows how he 'rationalizes the irrational' by constructing an underlying pattern to unite 'fact' and 'fiction'. This pattern – the part to which the whole can be reduced – is language or more broadly symbol. Through symbolic interpretation, Moore interweaves a range of beliefs in a pluralistic cosmology that draws heavily on traditions of western mysticism, especially the Kabbalah, but incorporating the Tarot, astrology, alchemy, Gnosticism and various non-western teachings and mythologies. He also incorporates science at a number of levels. In part, this repeats the structural continuity of superhero universes, but Moore goes further in seeking to construct a transcendent understanding that integrates science with these other traditions into a single, coherent cosmological order that defines the basis of reality itself. Language is fundamental, because it is through this that we communicate with a separate 'magical' level of reality, the world of ideas and imagination. In this neo-Platonic realm that Moore calls 'the Immateria, consciousness imagined as space' (Moore, Williams and Klein 2005: 7), the gods and all 'fictional' creations are real, each representing some aspect or dimension of 'Ideaspace'. Through symbolic mediation, our contact with these other-worldly Ideational entities is put into communicative and hence material form. In this way, we actively construct material reality – and that, says Moore, is magic.

In *Promethea*, this worldview is set out both discursively and visually (Locke 2008, 2009a), reflecting Moore's view of comics as itself a kind of magical medium where visual and textual symbols intersect and mutually play off each other. In comics, pictures become words and words become pictures (see also McCloud 1993, Varnum and Gibbons 2001); both become symbols, enabling a level of reflexive interplay that for Moore resembles magical consciousness and bears comparability to forms of ritual enactment found in both psychedelia and 'conventional religion' (Moore, Williams and Klein 2005: 20), where techniques of 'sensory overload' are used to access altered states of consciousness. One form of such symbolic interplay in *Promethea* is the use of metaphor to integrate magic and science. This is done in a deeply self-referential manner, incorporating both the form and content of the comic, most obviously through imagery of circles and spirals epitomized in the recursive circularity found in such mathematical symbols as the Möbius strip (Moore et al. 2001b: 8-9). Throughout the series, Moore draws analogies with other circular symbols taken from both science and magic including: the number eight (Moore et al. 2001b: 6-7); the symbol of infinity (Moore et al. 2001b: 8-9); the double-helix of DNA (Moore et al. 2001: 23) also represented by the snakes on Promethea's caduceus (see below) (Moore et al. 2000a: 15, 2001c: 4-5); the mystical symbols of the Ouroboros (Moore et al. 2001: 24) and the Kundalini (Moore et al. 2000a: 16-17); and the Tarot image of 'The Universe' as envisaged by Aleister Crowley showing the female moon dancing with the earthly snake – DNA again (Moore et al. 2001: 23, 2001a: 18-19). Thus, in this central feature there is recurring syncretizing of modern science and occultic lore through mutually informing metaphorical interpretation: DNA is interpreted through occultic serpentine imagery as much as the latter is interpreted through the former.

However, Moore makes clear where he puts priority, arguing that although 'science regards mind as an hallucination of our nervous system' ultimately reducible to 'hard physics', this itself 'rests on quantum mechanics, which Werner Heisenberg asserted as inseparable from the influence of the observing human mind' (Moore, Williams and Klein 2005: 21). Further, he argues that 'science ... is based on reliably repeatable phenomena, and therefore cannot discuss even everyday consciousness ... much less the exploding, psychedelic experience of magic, of our human psyche, or our soul'. He also represents this symbolically, depicting Promethea in conversation with the Tarot image of 'The Universe' asking: 'is the snake's head under the woman's foot because she's like, growing up out of his mind? Is his head there because material life has invented imagination?' To which comes the reply (from the sibilant snake): 'No. It'sss there becaussse I am her servant.' (Moore et al. 2001a: 21) Thus, much like Hubbard (see Chapter 5) and John Michell (see Chapter 7), Moore draws on a specific view of the nature of science as dealing with 'reliably repeatable phenomena' in order to present it as limited. Thus, although science is used as a resource to bolster the view of reality presented, the disenchanted viewpoint is also a focus of critique.

Circularity is also built into the form of one particular issue of the series, *Promethea* 12 (Moore et al. 2001), the separate pages of which fit together to make one continuous image in which the last page loops back to the first so the whole story may begin again. It could then be displayed as a recursive Möbius strip, much as the Tarot Major Arcana is depicted in some commentaries (for example, Douglas 1974) and the story sets out a cosmological narrative of the whole of creation from beginning to end, for which the individual human life is a metonym. Thus, the comic acts as a physical representation of Moore's conception of the totality of 'spacetime', a notion taken from science with the Big Bang at one end and a hypothetical 'Big Crunch' (Moore et al. 2004: 9) at the other, the whole seen as a single, eternal moment. Moore connects this to magic through interpreting the Kabbalistic 'Tree of Life', in a further circular technoscientific metaphor, as a 'circuit board' (Moore et al. 2001a: 8-9). Traditionally, the 'Tree of Life' is conceived as a lightning bolt descending from Kether, the Godhead, to Malkuth, the earth or material sphere, with an implied hierarchy of creation and unidirectional energy from God downwards. In Moore's revision, however, the energy flows as in a circuit only when the connection is completed, so replacing hierarchy with interdependence – neither Kether nor Malkuth exist without each other. Thus, the recursive loops he recurringly references are symbolic representations taken from scientific knowledge as well as magical lore of this infinitely circulating 'structure of things, whether that's the universe or each individual soul' (Moore et al. 2001a: 8-9), something *Promethea* 12 represents as both a physical and visual-discursive synecdoche.

Promethea as a character provides a similar cosmic connection linking humanity with the total cosmological order. Like other superheroes (Locke 2005), she enacts a mediating condition through her ordinary human identity, Sophie Bangs. Sophie ('wisdom') represents 'every(wo)man' and also magic, which is similarly female (Moore et al. 2000a: 18-19). She embodies the human potential to become transformed by and into magic, to become an enchanted self, an extraordinary otherness defined by paranormal abilities. This is signified by Promethea's costume, itself a syncretic construct drawing on Egyptian, Greek and Roman imagery representing the dual god of magic, Thoth-Hermes, and complete with caduceus, the two snakes of which represent the metaphorical spiral connections noted above. However, whereas for many superheroes, enchanted transformation is wrought by technoscientific means, for Sophie it is accomplished through poetry. Thus, the 'science-heroine' Promethea is channelled through the symbolic charms of language, the medium of magic. Promethea, then, represents consubstantiality connecting the human to the enchanted, magico-scientized cosmological order. Through her, human qualities are reconstructed in an enchanted-science form, just as enchanted science is also humanized.

This is apparent partly from the use of a metaphor of journeying, which is a standard trope of the priestly voice used to represent the transformation wrought by a transcendent understanding (Lessl 1985). Accordingly, Promethea explicitly undertakes a 'journey into magic' (Moore et al. 2000b: 22) which, as in traditional 'wonder-tales' (Propp 1968, 1984), represents the hero(ine)'s development to a new, higher condition, typically involving greater knowledge and self-understanding. Thus, through her, Moore uses the rhetoric of folklore and popular science to invite his readers into personal transformation to see reality itself as a magical construct. This is apparent in her relationship with the caducean snakes, Mack and Mike ('macrocosm' and 'microcosm' respectively) in which she takes the role of disciple to their teacher as her journey unfolds. Promethea asks her questions in Sophie's ordinary, everyday language while the snakes respond in poetry, just as Sophie uses poetry to become Promethea. Thus, Promethea stands to the snakes as Sophie does to her; and just as Sophie represents 'every(wo)man', so does Promethea represent us all as neophytes being inducted into the priestly order, an order at once occultic and scientific.

Moore legitimizes his enchanted cosmology by appeal to a notion of intelligent design, citing the 'strong anthropic principle' (Moore et al. 2004: 8). In a view reminiscent of Sagan's, he takes this as evidence that 'spacetime' is intended to support its own self-awareness in the form of human consciousness and it is through this that the human is scientized. Thus, in response to Promethea's questioning about magic, Mack and Mike invite her into 'the magic circus of the mind!' In speaking of '*the* mind', they perform a synecdochic reduction in which all minds are treated as one and the same; it is no particular mind or a special type of mind, but all and any minds that are cued. This is treated as the essential feature of human beings, which for Moore is also the site of 'Immateria'. However, he also gives it a scientific characterization through a description of the nature of 'intelligence' that, through quantification rhetoric (Potter, Wetherell and Chitty 1991), invites a sense of wonderment at its workings:

> Intelligence is not based upon the number of the brain's neurons, but how many CONNECTIONS [*sic*] we forge between them, linking one concept, one neuron with another, in a shimmering, self-aware neural net. The number of potential neural connections in the human brain exceeds the number of particles in the known universe. (Moore, Williams and Klein 2005: 3)

As with 'the mind', here it is '*the* human brain' presented as being an essentially identical, single thing, while in the metaphor of a 'neural net' (that 'shimmers') emphasis is given to the 'connections' rather than the unmentioned holes nets also have. The description is humanized through the collective pronoun 'we' as the active agent that does the 'forging' of these connections, themselves described as 'self-aware'. Given that this is said to be 'potentially' vaster than the universe itself, 'we' then exceed the cosmic order of 'spacetime'. 'We' can be omniscient; we are – each of us is – God.

Thus, an enchanted sense of self is drawn from a scientized description of 'the mind'. This is also given explicit linkage to the total cosmic order again described in scientific terms in a speech by Promethea:

> Our consciousness, a startling outgrowth of the universe, is possibly its most important part, the fraction of existence that can think, feel, marvel at itself. We are all spacetime's sensory organs, spacetime's mind, our thoughts and lives naught but the three-dimensional, material expression of its blazing, immortal soul. This jewel of being, Big Bang flared at one end, Big Crunch at the other, simultaneous, all going on right now, a perfect frozen fire. The world is young, our most remote ancestors not yet born. The world is old, and we have all been dead for decades, centuries. Don't you remember? (Moore et al 2004: 9)

Here, the consubstantial connectivity that scientizes the human is made explicit: it is 'we', synecdochically defined as 'minds' that constitute the point and purpose of the scientized cosmic order. The cosmos – scientifically constituted through enchanted imagery – and 'the human mind' – the seat of 'Immateria', the imaginary source of reality – are one. The final step is made by Promethea addressing us directly from the comics page, breaking 'the fourth wall', the boundary between 'fiction' and 'fact', to ask us: 'Don't you remember?' The implication being that all the knowledge and understanding behind the scientized cosmic order has always been known to us, which is why it can be found symbolically built into the images of the Tarot deck, the Kaballah, and all the other mystical and religious beliefs of humanity. They are all one symbolic order, multiple versions of the same transcendent reality, a single eternal pattern expressed in polyphonic heteroglossia but with one coherent pulse within. It is this coherent pulse that the priestly voice purports to chant, conveying to those of us without the hearing aid just what transcendent pattern beats behind the multiplicity of particular events. To do so, the priestly voice necessarily employs the available means of persuasion, the resources provisioned by the language, whether verbal or visual, present in the here and now, in this social and cultural context. The real magic of the priestly voice is then to pull off the trick of transcendence, when it is always rooted in the specificities and localities of particular times and particular spaces, inside history and society – and synecdoche and consubstantiality are its sleight-of-hand and smoke-and-mirrors.

Moore's version of the transcendent order, then, like those of creationism and Scientology, employs the rhetorical resources provided by enchanted science and occulture to construct a grander vision that purports to unite both in a single encompassing order. This is achieved through forging symbolic connections that are ultimately merged in an understanding of the nature of symbolization – language itself. In language, through symbol, we access and articulate the realm of the gods and this is as true of science as it is of mysticism. As suggested by Whitehead, this is the next level of the rationalization process in the further abstraction of the charismatic towards a more encompassing understanding. It is also, by a paradox of consequences, something that has occurred through the rationalization of the irrational, such that the discursive space constructed as a dumping ground for the rejects of science – and of them all, the thrice damned superhero! – is now employed as an arena in which to build pretenders to the scientific crown. Like so many hybrid cyborgs, the enchanted technoscientific Frankenstein monsters and golems of the sf imaginary are staging a fight back, turning on their scientific creator and threatening to remake reality in their own image. But it is important to see that there is nothing 'irrational' about this and that the 'rationalization' involved is entirely ordinary. Through the documentary method of interpretation, it is mundane reason at work, something to be considered further in Part III.

PART III
Mundane Mysteries

Chapter 7
Paradoxes of Definitiveness: Mundane Mysteries and Fortean Strangeness

Part II focused on some forms of enchanted science found in contemporary culture to support the claim derived from re-crafting Weber's thesis that rationalization does not necessarily entail a monolithical condition of disenchantment in the modern mental outlook. Rather the process of rationalization, understood as the abstraction of the charismatic into a more encompassing transcendent discourse partly fed by scientists themselves employing enchanted rhetorics in their public self-presentation, has produced a rich diversity of such enchanted reachings seeking to syncretize science with religious teachings and/or magical beliefs. Through a paradox of consequences, far from eroding moral and mystical outlooks, the rhetoric of science has contributed to the development of an enchanted discursive multiplicity in which it is employed both as a means to legitimize traditional moral commitments and construct alternative mysticisms. Alongside this, however, stands another paradox of consequences in which disenchantment itself leads not to a condition of determinate knowledge but a state of unresolved mystery. The belief in the in principle knowability of the world has encouraged efforts to eradicate mystery that have instead resulted in its elaboration and diversification, specifically in the form of a range and variety of activities and viewpoints on the 'fringe' of established science.

Parts of this fringe that together I call *mundane mysteries* are the focus of the next three chapters. A basic point here is that the discourse of the fringe arises from the workings of mundane reason in its disenchanted condition and this needs some explanation before turning to specific cases. A key focus is the discourse of the followers of the early twentieth century American 'philosopher' (their term), Charles Fort. Fort collected what he (1995) called 'the damned', that is 'data' in the form of reports of various phenomena that were ignored, overlooked or discounted by established science and often presented ostensible challenges to its claims about the nature of reality. Many of these reports were taken from scientific journals and contemporary Fortean interests continue to encompass much of fringe science out of which their discourse partly arises. Thus, Fortean discourse is generated from the workings of mundane mystery directed toward 'damned' fields such as ufology (the study of unidentified flying objects), cryptozoology (the study of anomalous life-forms), parapsychology, and alternative archaeology. However, in keeping with Fort's scepticism towards any authoritative beliefs including science – and even his own – Fortean discourse also ranges beyond fringe science resulting in internal divisions that reproduce those found in the discourse of scientists in respect of accounts of error. Thus, Fortean discourse mirrors scientific discourse albeit in a 'distorted' fashion as seen from the perspective of formal science – although from the Fortean viewpoint this perspective is itself 'distorted' by virtue of its self-limiting reductionism that excludes the 'damned'. Forteanism, then, explicitly tolerates reality disjunctures leading to the generation of a more encompassing conception that seeks to account for the unaccountable. As such, Fortean discourse has some comparability to the transcendent frameworks considered in Part II, but differs in that it is not unified and, at least in some versions, delights in its own incoherence.

Out of the logic of accounting for error arises one particular branch of Fortean interests that is intriguingly troubling for sociology, conspiracy theorizing – or what, at the risk of over-using the word, I shall call conspiracy discourse. Conspiracy discourse is a form of 'social and cultural theory' (Bell and Bennion-Nixon 2001: 49) that has apparently enjoyed growing popularity in recent times (Knight 2000, Parish and Parker 2001, Barkun 2003), but its ostensible sociological form ironizes standard sociological accounts. I consider this in Chapter 8, where I argue it is possible to construct a sociological account in the terms of re-crafted rationalization. Finally, in Chapter 9, I look at one further topic of Fortean interest, Jack the Ripper, to begin to consider what re-crafted rationalization has to say about the membership category of the 'scientist'.

Mundane Mysteries

First, however, to explain the term 'mundane mysteries', which I use to highlight some basic similarities of feature between disenchantment and Pollner's (1987) concept of 'mundane reason'. The starting point here is that disenchantment entails an interest in 'mysteries' and thence enables matters to be designatable as such by creating this discursive space. This is not to say that 'mysteries' did not exist prior to the disenchanted outlook, but that it encourages a particular orientation or directing interest toward them with the specific intention of dissolving their mystery. This follows from Weber's view of disenchantment as the belief that the world is knowable in principle and so is intolerant of the 'unknown'. Although, as Weber (1948b: 139) states, this need not necessarily lead to any particular attempt to make the unknown known – the man or woman on the Clapham omnibus may feel no particular urge to learn about its mechanics – but only to the conviction that it is knowable, it does entail an in principle refutation of any alternative belief to the contrary. Thus, under the stimulus of the world-tranformative motivation inherited from Protestantism (as discussed in Chapter 2), the disenchanted orientation has developed the institutional means intended to prove its point in the form of modern science – a range of organized activities directed towards the demonstration that what was taken to be unknowable can in fact be known broadly through the 'experimental way of life' (Shapin and Schaffer 1985). Thus, as epitomized in science, disenchanted discourse takes it as given that the world is determinate. Further, as discussed in Chapter 3, this also encourages intolerance towards contradiction as expressed in the rhetoric of rationalism. In addition, science pursues a fully coherent description of the totality of reality, as seen in unified field theory, 'a single theory that describes the whole universe' (Hawking 1988: 10). In Pollner's (1987: 41, quoting Merleau-Ponty) terms, such a theory has as both its inspiration and intention, the 'Great Object': 'the world, as the over-arching context of lesser objects and thus an object itself, is idealized as a "finished explicit totality in which the relations are those of reciprocal determination".'

For Pollner (1987: 17), the Great Object is an 'idealization' of mundane reason that follows a set of assumptions that the world is 'determinate, coherent and non-contradictory' – just the features noted above about disenchantment. Viewed in this light, then, disenchantment shares the same set of presuppositions as mundane reason, so they have something of an 'elective affinity' (Weber 1948c: 284). For Pollner (1987: 56), mundane idealizations constitute 'incorrigible proposition[s]' that cannot be proven false under any circumstances, because they are the means whereby descriptions of reality are constituted and assessed. Mundane reason takes it as given that reality has these qualities, so if claims are made that conflict with them, then so much the worse for the claims; the idealizations will not be questioned because they are used as the basis to assess validity. In this way, mundane reason is preserved from 'reality disjunctures', ostensibly conflicting accounts of objects in the world or experiences of them. In the terms used earlier with reference to science, mundane reason has a set of resources for accounting for error, ways of calling into question the status of statements, objects or experiences in order to reconcile them with the pre-given assumption that the world is determinate, coherent and non-contradictory. Thus, there is further commonality with the discourse of disenchantment – the formal rhetoric of science – which has generated its own specific standardized verbal formulations to account for error.

Perhaps, then, disenchantment and mundane reason are simply the same thing? One question here concerns the status of mundane reason in terms of its degree of generality. Pollner's discussion mainly uses examples from traffic courts in the United States, which although a good site of conflicting accounts and the techniques used to resolve them, leaves some uncertainty as to the broader position of mundane reason. At times, he seems to treat it as constituting the general commonsense outlook of modern western societies, especially because he associates it with the subject-object divide. Thus, he (1987: 129-147) strikes particular contrast with social worlds that seem to lack this such as Zen Buddhism. However, that he does not see mundane reason as solely limited to the modern west is apparent from his (1987: 53-58) discussion of Azande witchcraft beliefs. On the other hand, he (1987: 77-78) also recognizes that even in western commonsense, reality disjunctures are not always treated as problematic. Thus, he (1987: 58) quotes Pirandello's remark, 'it is so ... if you think it is so', which I take as some indication that western commonsense has amongst its commonplaces those that enable the acceptance, or at least deferral, of conflicting accounts of reality, without the felt necessity to resolve them in accord with mundane idealizations – as in the everyday expression, 'it's all a matter of opinion'. Given this, I suggest that mundane reason be viewed as a discourse that has become strongly institutionalized in some contexts in western societies in the modern period, but that its extension beyond these is an open matter.

Here, the parallel with disenchantment becomes even more marked. As I have argued in preceding chapters, disenchantment should be viewed as a discourse that constitutes part of the formal rhetoric of science employed for the purposes of some public self-presentation. This discourse bears marked parallels with mundane reason: they share similar idealizations and similar resources of accounting for error. However, just as there is reason to doubt that disenchantment adequately describes the mental state of the ordinary person as opposed to being an occasioned set of rhetorical resources, so too is there reason to suggest the same for mundane reason. Thus, just as some people may some of the time argue against disenchantment, so too may some people sometimes argue against mundane idealizations. This is what can be seen in fringe science and in Fortean discourse more especially.

The matter can be turned around to ask the question of what happens to mundane reason under conditions of disenchantment. Two things in particular arise: anomalies and mysteries. From the point of view of disenchanted mundane reason, in general these are both measures of indeterminacy, ways of describing as yet unknown knowables – the 'known unknowns' as it were. However, whereas 'anomalies' are generated internally within science as an outcome of its procedures of enquiry (Kuhn 1970), 'mysteries' are generated externally as phenomena, experiences or statements the claim to reality of which is potentially discountable by the scientific community by virtue of their non-scientific sourcing.

Two points come from this. Firstly, rather than being able to bring about its Great Object, the disenchanted pursuit of knowability itself generates unknowns specifically through the activity of scientists. At one level, the procedures of empirical enquiry themselves continually generate new phenomena as potential candidates for factual status that, if they are to be so accorded, require incorporation into the existing Great Object. Problems can arise, however, because as Pollner (1987: 37) argues, 'definite relations of continuance, complementarity and conformance among aspects [of the Great Object] are anticipated'. That is, 'aspects' – specific 'objects' or 'facts' – must not merely fit in, but fit 'harmoniously' and in a suitable fashion otherwise disjunctures will be apparent. Thus, it is not just the production of candidate 'facts', but their interpretation that can produce disjunctures, with resultant controversies that can incorporate both the 'factual' status of phenomena and their interpretative fit with existing frames of meaning (Collins and Pinch 1982; see also Billig 1996: 93-111).

The resolution of such controversies is far from guaranteed (Collins and Pinch 1993) and Pollner (1987: 76-77) points to the potential for a similar vicious circle in mundane reasoning as Collins (1985) identifies in the 'experimenter's regress'. The regress arises from a fundamental conundrum at the heart of the experimental way of life: an experiment is undertaken to find the right answer, but we cannot know if the experiment has been done correctly unless we know what the right answer is. As a consequence, controversies can potentially go on forever in a continuous spiral of charge and counter charge over the adequacy of methods and the validity of findings, as the case of parapsychological research exemplifies (Collins and Pinch 1979, 1982, Wooffitt 1992, 2006, Hess 1993). Pollner (1987: 77) likewise refers to an inherent uncertainty in the relation between individual empirical experience and intersubjective reportage of such experience (the instance involves two males): 'equivocality is perpetuated by a mode of reasoning in which each of the disputants, treating his version as a given and thereby ironicizing competing experiences, finds the experiential claims of the other to be the product of an inadequate procedure for perceiving the world.' Thus, reality disjunctures can potentially be upheld indefinitely *because of* the mundane idealizations that the world is determinate and coherent. These idealizations insist that the world is not merely knowable but knowable in common, that it is the same world for all of us and therefore that our experiential descriptions should be in agreement as 'harmonious' aspects of the prevailing Great Object. However, because different versions of events are possible and the same warrants and discountings are available to each, disjunctures can potentially be maintained indefinitely.

Disenchantment, then, despite and because of its assumption of the in principle knowability of the world, contributes to the production of unknowns, some of which at least persist even if in a marginalized position on the fringes of scientific orthodoxy (Wallis 1979). From here, they may be taken up to become mysteries, something that can be facilitated by fringe scientists themselves seeking popular publicity and support once denied orthodox acceptance (for example, Brossard 2009). This points to what Pollner (1987: 70) calls the 'politics of experience'. Marginalization is one strategy in the attempt to enforce resolution of reality disjunctures by imposing one version as definitive. However, it can backfire and contribute to the expansion and perpetuation of disjunctures between the orthodox scientific view and those circulating amongst the wider public, since the presence of fringe scientists can provide a resource to legitimize forms of 'deviant' knowledge. Thus, scientific expertize can be mobilized 'tactically' (Nelkin 1992b) as a means to support or legitimize knowledge-claims and to undermine or delegitimize others, including those of the orthodox scientific community.

One way this tension can be exacerbated brings in the second point. Whilst science produces anomalies, activities external to science can produce mysteries, that is events and phenomena construed as such as a result of a disenchanted worldview configuring commonsense reasoning in its terms. Clearly enough, the construal of events and phenomena as 'mysteries' is relative to a surrounding sense of reality as demystified; the 'strange' is only such against a background horizon of expectations taken as 'not strange'. Under disenchantment, however, 'strangeness' becomes the focus of a specific kind of puzzlement, an unknown that at least potentially can be made known, rather than simply be accepted as the whim of the gods, a boundary not to be breached or as just a feature of the landscape.

One example comes from a book by John Michell (1973) that was inspirational to many contemporary Forteans (Rickard et al. 2009). Regarding the 'discovery' in 1648 by John Aubrey of the megalithic 'prehistoric temple' at Avebury in Wiltshire, Michell (1973: 1) wrote: 'It was not hidden in some remote and desolate spot, for a thriving village stood within its ramparts, nor at that date was it particularly ruinous. Yet Aubrey was the first of his age to notice it.' Since Aubrey, successive generations have viewed the standing stones and their arrangement as a mystery to be solved using the available means, instruments and reasoning of the mundane idealizations of their time. In recent decades, this includes many like Michell himself, who lack 'expertize' in the sense of formal scientific qualifications, but for whom 'mysteries' like Avebury are a focus of dedicated interest and extensive speculation (for example, Devereux 1991). This has resulted in what seems to have been a considerable expansion in the number and range of mysteries (Wilson 1979, Rickard and Michell 2000) to which Fort (1995, 1996, 1997, 1998) himself was a significant contributor (Steinmeyer 2008).

This expansion follows from disenchanted enquiry 'discovering' mysteries. Of note here is Michell's (1973: 4) comment about local farmers in Aubrey's day digging up standing stones for use as building materials, which might be viewed as a rather more 'instrumental' outlook than the abstracted puzzlement that cloaks them in an aura of mystery (see also Letcher 2006). This points to the disjuncture between 'calculation of means' and 'theoretical mastery' discussed in earlier chapters. Viewed rhetorically, the contrast structure (Smith 1978) works to accentuate the 'strangeness' of Aubrey's way of seeing, sharpening the puzzle of his 'discovery'. Discovery accounts are often constructed as journeys into the unknown (Brown 1994), the outcome of which is a new, higher state of knowledge or understanding that, as discussed in earlier chapters, accords with charismatic reaching. As in the spirit of positivism, then, the disenchanted promise of demystification paradoxically generates a promise of greater enchantment and this contributes to the tensions informing mundane mysteries. For while the disenchanted sensibility turns perceived and reported phenomena into mysteries to be enquired into and solved, the issue arises of what specific kind and form of solution is to be advanced. In Pollner's terms, what is the 'Great Object' with which solutions must be in harmony? Here, tensions with science may be exacerbated as, while for some the Great Object is that given by prevailing scientific orthodoxy, for others such as Michell, it is not. Like Hubbard (Chapter 5) and Moore (Chapter 6), Michell (1973: 4) distinguishes 'two different forms of science', defending a neo-Platonic view against what he (1973: 19) calls 'the mechanistic approach'. Thus, the tensions and irresolutions within science provide resources for a counter-politics of experience, in which the means of accounting for error come critically into play.

As discussed in previous chapters, scientists have available a repertoire of means of accounting for apparent discrepancies in reports of phenomena as due to errors I have categorized as methodological, personal and sociological (Pollner gives a somewhat different set of resources for accounting for reality disjunctures, but the discrepancies are relatively minor). Such accounts of error also feed the fringe, with the consequence that it has not only grown but thickened. Thus, rather than disenchanted mundane reasoning producing a single definitive version of reality, a multiplicity of versions have been generated out of the apparent reality disjunctures presented by anomalies and mysteries, inspiring competing accounts seeking to make the unknown known, but doing so through contrasting versions of the 'known'. In addition, the identification of errors as due to methodological, personal or sociological factors leads to the production of would-be explanatory understandings of the reasons for such errors, producing methodological and philosophical debates, and psychologies and sociologies of error. These also feed the fringe, providing resources of argumentation to defend preferred and undermine dispreferred versions. All of this can be seen in Fortean discourse, as also can the further feature of the toleration of reality disjunctures in a form of reasoning that is both non-disenchanted and non-mundane but that still persists within modernity.

Fortean Discourse

What follows is not an exhaustive account of Fortean discourse, but only intended to provide illustration of the preceding points. The primary source used is *Fortean Times* (FT), a monthly magazine first published in November 1973 that provides the main outlet for Fortean topics in Britain. Its content chiefly consists of news reports, articles and readers' letters and I have drawn from issues published over the preceding five years at the time of writing, from December 2004 (FT192) to November 2009 (FT255). The first point is to establish that Fortean discourse is constructed around a sense of the mysterious and for this no more is needed than to refer to FT's sub-title, 'the world of strange phenomena'. Thus, the content is pre-figured in a manner that authorizes a version of the content (Smith 1978) as a documentation of 'strange' stuff. 'Strangeness' is how the 'out-there-ness' (Potter 1996: 150-175) of 'the world' is constituted. Nonetheless, Fortean discourse is informed by mundane idealizations, viewing reality as a determinate, coherent and non-contradictory Great Object and accordingly, it employs mundane self-preservations built around accounts of error, three prominent types of which refer to: *technologies of detection; psychologies of perception;* and *sociologies of conception*. These are used in a politics of experience that involves the construction of two broad types of membership category, externally between Forteans and non-Forteans, and internally amongst Forteans. In a manner that parallels orthodox scientists, Forteans distinguish themselves externally as 'experts' from 'non-expert' others, including both scientists and the wider public, and internally through forms and degrees of 'critical' reasoning generally referred to as 'scepticism'. However, although these techniques are employed to resolve reality disjunctures by ironizing knowledge-claims, Fortean discourse often tolerates disjunctures and even presents these as anticipated confirmations of the Fortean Great Object, sometimes represented as a Trickster, the determinacy, coherence and non-contradiction of which resides precisely in the absence of these qualities.

Mundane Idealizations

An example of mundane idealizations at work in Fortean discourse comes from FT's review of reports made in late 1980 by US Air Force personnel stationed near Rendlesham Forest in East Anglia, of encounters with unidentified flying objects (UFOs) partly in the form of unusual lights in the sky (Randles and Clarke 2005). One source of controversy regarding reports of UFOs is that they are seen by many scientists as conflicting with mundane idealizations that reality is determinate, coherent and non-contradictory. This is especially so if it is assumed that the term 'UFO' refers specifically to spacecraft ('flying saucers') piloted by alien beings – the so-called 'extra-terrestrial hypothesis' (ETH) (Denzler 2003; for an account of mundane reasoning in a 'flying-saucer group', see Tumminia 1998). As mentioned in Chapter 1, the possibility of life on other planets is itself disputed, but far more controversial is the idea that any such beings might be visiting earth, as this seems to conflict with established scientific knowledge about the possibility of intelligent life elsewhere in our solar system or travelling across interstellar space. Thus, for 'flying saucers' to be visiting earth would seem to threaten the coherence of the scientific Great Object. Accordingly, reports of encounters with UFOs are typically subject to 'debunking' (which I should make clear is not my own intention regarding Forteanism). Debunking attempts to maintain mundane idealizations by removing the apparent contradiction with (at least one version of) the Great Object. One strategy routinely followed is to look for possible 'known' objects that might account for the reported experience, so making the object no longer *un*identified but determinately identified (Westrum 1977, Dean 1998: 34-38, Denzler 2003). In the Rendlesham case, one such object was the Orfordness lighthouse, which was made a point of specific interest by one FT writer: 'Once the press attention on the case brought forth the idea that the Orfordness lighthouse might be responsible for the "UFO" seen through the trees, I went straight to Rendlesham to check out the theory.' (Randles 2005: 38)

A key feature of mundane reasoning is to treat all aspects of any given object as in conformity with each other; for it to be the self-same object, it is assumed to display a self-consistency observable to anybody looking. Moreover, any given description of the object is assumed to be open to potentially 'infinite explication ... which at once promises and is informed by a horizon of further determinations which, if known, would render the intended object complete' (Pollner 1987: 33-34). Thus, much like comic-book continuity as discussed in Chapter 6, descriptions of real world objects are assumed to be open to potential 'filling in' using the documentary method of interpretation. Ultimately, this extends to the Great Object as a whole, so that self-consistency applies not only to the object but to the context in which it is embedded. Thus, mundane idealizations apply to the 'object-in-context' which is treated as just another object, as is '(the object-in-context)-in-context' (Pollner 1987: 41) and so on up to and including the whole of reality.

This is apparent in Randles's description of her action in 'checking out the theory' that the Orfordness lighthouse might have been 'responsible' for the reported lights. Her action prospectively treats both the original reported object (the lights in the sky) and the object-in-context (the light in relation to 'the lighthouse ... seen through the trees') as determinate, coherent and non-contradictory. For the theory to be checkable in the manner she describes presumes that the world observable to her is essentially the same as that observable to the airforce personnel, and that it is determinately so through the activity of observing it. It also assumes a consistency between the light they might have seen and the light visible from the lighthouse through the trees, between, that is, the object and the object-in-context embedded in specific features of the surrounding landscape. The selection of these specific features is important in the construction of the reality described, since a potentially infinite number of features might be chosen. Randles's selection is attributed to the 'theory', but in principle any and all other features could be considered to 'check' the consistency of the reality observed.

This is apparent from her (2005: 38-39) subsequent comments: 'there is a problem with the lighthouse theory as the "big answer" to the case, because the locals all knew it was there and saw it night after night. There were many local witnesses to strange things that weekend – not just nervous airmen in a dark, unfamiliar forest.' Here, she now locates the object-in-context (the lighthouse seen through the woods) in a further context, 'local' knowledge. Thus, the object-in-context is now treated as an '(object-in-context)-in-context' in order to 'check' the theory at another level. This again is in accord with mundane idealizations that this larger object should be non-contradictory and it is this assumption that produces the 'problem'. To be self-consistent, this larger object would have to conform with the aspects of it Randles highlights as significant, locals' knowledge and regular experience of the lighthouse. But here she finds an inconsistency and so a 'problem' is generated from the disjuncture between the specified features of the larger object and the mundane idealizations. Thus, as Pollner argues, under such circumstances, it is the idealizations that are preserved not the imputed reality – the 'theory'. Hence, the theory can be adjudged inadequate and the 'strangeness' of the original object, the lights in the sky, sustained.

Mundane Self-preservations

Also notable is that Randles provides a warrant for her selection of the specific aspect of the larger object by reference to 'local witnesses to strange things'. Again, mundane idealizations are involved in the assumption that the objects – 'strange things' – these locals 'witnessed' would have been the same as those seen by the airmen. However, in being described as 'local', they are contrasted with the airmen described as 'unfamiliar' with the vicinity. This invests the locals with greater credibility as observers; they appear not only as relatively knowledgeable by virtue of this 'location formulation' (Silverman 1998: 134), but also as more objective, as sure 'witnesses' in comparison to the airmen whose 'nervousness' exacerbated by the 'dark, unfamiliar forest' (no longer just 'trees') might have adversely affected their capacity to observe reality correctly. (There is an obvious, if ironic parallel to be drawn here with the upgrading of 'local knowledge' in PUS – for example, Wynne 1992.) Here, then, we see a form of mundane self-preservation prospectively accounting for error and used to sustain the reality of the 'strange things' as such.

However, as argued above, accounts of error can cut in both directions and this is common in Fortean discourse where contrasting versions are in turn undermined and sustained through accounts of error advanced by contending sides. Three common forms of error accounting in Fortean discourse refer to technologies of detection, psychologies of perception and sociologies of conception. These parallel forms of contingency used by scientists in controversy and they also draw on developments from scientific studies of error, concerned to document and explain their existence in a manner taken to be consistent with the scientific Great Object. Thus, in the first case, whilst in scientists' discourse a key focus of controversy concerns the adequacy of methods employed in empirical research, for Forteans a more common equivalent concerns the means employed to detect 'strange' phenomena. Often, technological means of detection are employed, such as forms of photographic and video equipment, which effectively function as intermediary observational devices taken to be extensions of humanity's own sensory capacity. However, just as experimental methods may be made the basis for questioning laboratory observations whether made directly by people or indirectly by machines (Latour and Woolgar 1986), so may intermediary observational devices be subject to similar questioning.

A number of questions concerning the status of technologically detected objects are possible: are they the same objects; are they objects at all; and are they the objects they appear or are claimed to be. An example of the first concerns a 'star map' (Holman 2008: 50) originally drawn by an American woman, Betty Hill, who claimed to have been abducted by aliens in 1961, during which she copied the map from a chart of 'trade routes' used by her abductors (see also Dean 1998: 48-50). According to Holman, in the mid 1960s an attempt was made to compare the map with a model constructed from the position of known stars based on currently accepted astronomical data, which he (2008: 51) describes as 'not very convincing'. However, in 1972, revised data were published 'refining and in some cases altering considerably the distance measurements' for some stars and this produced a better fit, which was taken by some as evidence to support Hill's claims. However, star measurements have since been revised again with information provided by new technologies of detection: 'For example, in the early 1990s the Hipparcos satellite measured the position of nearly 120,000 stars 10 times more accurately than ever before' (Holman 2008: 51). As a result, several stars in the model now appear closer together than earlier measurements indicated and Holman takes this to undermine the credibility of the map. Here, then, the production of new measurements by technologies of detection is not taken to be evidence that the world has changed, that the stars are not the same stars, but rather that the original measurements were wrong – even though they also came from then available technologies of detection. Thus, mundane idealizations concerning the consistency of reality are used to cast doubt on the adequacy of intermediary observations, even though the revised version of reality is itself only available through such means. The new measurements are viewed as 'more accurate', as closer to the way the world is and thus the new technology taken to provide a truer picture than older technologies (for a comparable instance concerning the detection of the 'hole' in the ozone layer, see Grundmann and Cavaillé 2000; see also Messeri 2010).

Technology, then, is presented on one hand as a pure medium of transmission of information, but on the other as distorting information, which raises the second way of questioning the status of objects by asking whether they are 'objects' at all. A topic that has generated some interest amongst both Forteans and a wider public over recent years is 'orbs'. These are 'balls of light, varying in size and usually orange, red, yellow or white in colour' (Anon 2007: 73) that frequently appear in photographs taken with digital cameras. They are often interpreted as evidence of some kind of strange phenomenon, 'most commonly ... the souls of the deceased'. Unusually on this matter, FT's editors (Anon 2007: 73) expressed an unequivocal view that orbs 'can be comprehensively explained ... [as] nothing more exciting than airborne particles ... between the photographers and their subjects.' Such particles can reflect the light from a flash 'and vary in apparent size according to their shape and distance from the camera'. Thus, the status of 'orbs' as genuine objects in the world, specifically as 'balls of light' is called into doubt by virtue of the light being interpreted as a technological artefact, a product of the means of technological detection detecting itself. Again, then, the method used to illuminate (literally!) the world is held to be responsible for the phenomena produced. As in the case of the experimental methods of science, questioning the adequacy of the method as a means of detecting reality is always possible, so it can always be claimed that 'objects' are not 'facts' but 'artefacts'.

This shades into the third type of questioning, which asks whether the object is what it appears or is claimed as being, leading to further questions about the nature of human perception and human motivation respectively. Where intermediary observational devices are involved, the capacity to postulate misperception or misdirection is ever present. Here, the regressive conundrum appears again, as while photographic or filmic evidence is often held out as the long hoped for final proof of a strange phenomenon, the very fact of technological mediation can be cause for doubt about the credibility of the evidence presented. Thus, the question of 'fact' or 'artefact' is especially prominent, giving rise to a particularly contentious feature of Fortean discourse, hoaxing (Brookesmith and Irving 2009a, 2009b, 2009c). Accusations of hoaxing are predicated on mundane idealizations in a manner that parallels Randles's attempt to 'check out' the visibility of the Orfordness lighthouse, which assumes the same activity by different people will produce the same observations (comparable to replication in scientific study and the associated assumption of standardized laboratory procedures in the empiricist repertoire – see Chapter 3). In the case of hoaxing, a sort of reverse corollary is at work to the effect that, if the same kind of technological product such as a film can be artifically made by another person, then the original may also have been artefactual.

An example of this concerns the 'Patterson-Gimlin film' (Korff 2005) of Bigfoot, the purported Sasquatch-like 'man-beast' (Coleman 2004: 58) said to inhabit the forested mountains of North America. The existence of these and similar creatures like the Yeti is disputed partly because it conflicts with the scientific Great Object that assumes all such hominids apart from *Homo sapiens* are extinct. Instead, stories of man-beasts are viewed as mythical inventions of indigenous folklore (a major source of cryptozoological interests – Magin 2008). Thus, the film is seen as highly significant and has been described as 'the most convincing piece of evidence yet to emerge indicating that Sasquatches are a biological reality' (Perez 2005: 36). However, it has also been described as a proven 'hoax' (Korff 2005: 34). In support of this contention, Korff (2005: 37-38) refers to 'confessions' made by those involved in its making and 'other key eyewitnesses', but adds that, '[t]he best way [to expose the truth] ... was to recreate the famous hoax film'. Thus, despite the existence of 'confessions' and 'eyewitnesses', it is in the fact that a comparable film can be made that the 'truth' is 'best exposed'. By demonstrating the possibility of artefacticity, the possibility of facticity, of 'biological reality' is undermined. If a technologically mediated observation can be artificially replicated, then its existence as a phenomenon in the world is cast into doubt.

This is indicative of a broader dilemma in our relationship to technology as to what role it should have in the social world and how far can it be trusted (Weizenbaum 1984, Woolgar 2002b, Locke 2005). Both sides in this controversy represent technology as trustworthy, but while one trusts it to present 'reality', the other trusts it to 'expose' this as unreal. Thus, although technology promises to resolve the paradox of mundane reason it actually reproduces it. As Pollner (1987: 26) states: 'When mundane inquiry reaches out for the "real", it is confronted by a paradox: the real is precisely that which is independent of its "grasp", and yet it is available only through some sort of grasping.' Thus, technologies of detection are assumed to be unequivocal mediators of reality and the idealizations attributed to objects – as determinate, coherent and non-contradictory – assumed also to apply to them. However, technologies are found not to resolve the paradox, merely to reproduce it at one remove. Independent 'facts' remain unavailable and the grasp of technology no more unequivocal than the people using it for grasping.

The equivocalness of people is at the centre of the two other means of accounting for error. Whilst the status of objects may be questioned, a further means of mundane self-preservation is to question the status of obervation as an experience. A rich variety of possible grounds for questioning observers' experiences are available to mundane reasoners and, as seen in the example above concerning the 'airmen' at Rendlesham, these may refer to emotional states as well as knowledge. This example is also interesting because it reverses what is often used as a means of bolstering the status of observers' experiences by reference to their special qualifications as observers. In UFO sightings, reports from 'trained observers', such as pilots and military personnel, are often presented as more credible than those of ordinary people (Dean 1998: 44-45). In the terms introduced in Chapter 1, these membership categories have associated with them the category-bound activity (CBA) of observing. However, Randles mobilizes the alternative membership category of 'locals' with whom she also associates the same CBA relative to the 'airmen'. Thus, membership categories can be used in highly flexible ways as is discussed further in Chapter 9 with respect to the category of 'scientist'.

The grounds for questioning obervers' experiences has been added to through the growth of psychological study of human perception with the additional caché of scientific legitimacy. In respect of alien abduction, for example, one commonsense resource that might be used to question such experiential claims is to say that someone was dreaming or seeing things that were not there. However, in the logic of disenchantment, such an explanation advances a further unknown to be puzzled over regarding the nature of these conditions and how they might affect 'normal' perception. Moreover, mundane self-preservations need to be in harmony with the scientific Great Object. One candidate explanation that fills these requirements is 'hallucination' (Pollner 1987: 61), which attributes the experiencing of non-existent things to a form of aberrant psychological condition affecting perception. However, this generates another unknown about what causes hallucinations, stimulating further psychological enquiry to provide candidate explanations in accord with scientific mundane idealizations. One FT letter-writer describes some such explanations as 'hypnagogic and hynopompic sleep states', that is 'falling asleep or waking up' in which 'hallucinations' may be experienced because 'the "reality" part of the brain is no longer working' (Gordon 2009: 75). Similarly, another condition called 'sleep paralysis' is 'often accompanied by auditory, olfactory and visual hallucinations as well as the sense of a "presence" nearby', all experiences that accord with reports given by abductees. In other words, they were dreaming – but at least it is now known they did so in a scientific idiom in harmony with the appropriate Great Object.

However, although such explanations may provide a suitably 'scientific' understanding of the general possibility of seeing things that are not there, they do not account for the specific claim that it is *aliens* that did the abducting. Here, we move from the psychology of perception to the sociology of conception. One way in which sociological reasoning is used in Fortean discourse is to present a view of the mental state of individuals as shaped by their social or cultural context in relation to one or other specific 'strange' phenomenon. So, one explanation of 'why aliens?' makes a similar point to Whitehead (1974) regarding science fiction (see Chapter 6) that the characters and motifs from the past have been translated into the scientized idiom of the present, noting a shared narrative form between contemporary accounts of alien abduction and traditional folk-tales about 'kidnappings by fairies' (Rickard and Michell 2000: 142). A more specific example is the explanation proposed by Roberts (2004: 37-38) of 'Midlands housewife' Cynthia Appleton's claimed encounters with a Venusian in the late 1950s. Discussing the question of a possible hoax on her part, Roberts points to features of the contemporary culture including that 'newspapers and magazines were full of' reports about flying saucers. Thus, he argues that 'even if Appleton's amazing tales *were* hoaxed, their significance is that they were triggered and generated by popular flying saucer culture, and may have held a reality for her.' Popular culture, then, is used as an explanatory resource in a manner that attempts to save the 'reality' of Appleton's experience, while also suggesting a cause for it outside of the experience itself and one that is at least prospectively consistent with the scientific Great Object (if one ignores the reality disjunctures between sociological and other scientific views of human action anyway!). I return to the blaming of culture in the next chapter.

Other kinds of sociological account appearing in FT are less accommodating of the 'reality' of the experience, though they too see the mass media as a significant causal factor, as in the supposed phenomenon of 'mass hysteria' (Rickard 2009). Pollner (1987: 82) discusses 'hysterical contagion' as an illustration of the privileging of analysts' accounts of reality over that of the people involved (see also Westrum 1977). Like the notion of 'hallucination', it poses further unknowns, stimulating 'a search for the presumptive sociological ... mechanisms' behind the condition. One such mechanism often identified is the mass media. Here, we encounter again the same tension as seen with the Bigfoot film, but on a wider social scale: Forteans rely heavily on media reports for information about the 'strange phenomena' they document, but the media is also often a focus of Fortean criticism for its supposed inaccuracies and distortions (for example, Roberts and Clarke 2009b). Indeed, it is sometimes seen as more or less directly responsible for 'shared delusion and strange behaviour' (Rickard 2009: 30), as in the classic case of Orson Welles' radio play of *The War of the Worlds*, broadcast in the United States on the night before Halloween in 1938 (Bartholomew and Evans 2005). The play presented the story of Martians invading earth as though it were a live broadcast recounting events actually happening at the time. There has been some debate ever since as to the impact of the broadcast, between claims that it caused mass panic and counter-claims that its effects have been 'exaggerated' (Bartholomew and Evans 2005: 42; see also Cantril 2005). In arguing against the latter view, Bartholomew and Evans document contemporary press reports about the extent of the 'panic' and the form it took. They do this without ironic intent, taking the press reports at face value as direct representations of the reality of the situation. Where irony *is* found is that they (2005: 46) do so in order to make a case for the human 'capacity for self-deception', specifically in the form of 'seeing what it expects to see'. Thus, the thrust of their argument is that many people who heard the radio show suffered such 'self-deception'. The question, however, is *whose* self-deception is this? In treating media reports as direct representations of reality whilst arguing that people see what they expect to see – an argument made with direct reference to the media – Bartholomew and Evans nicely, if unintentionally, encapsulate the tension in Fortean discourse regarding the media. It is at one and the same time a source of information and a means of deception; it is both truth and lies – a situation that encapsulates also the quandary of mundane idealizations.

The Politics of Experience

'Truth and lies' could be taken to sum up the whole of Fortean discourse, especially in its most encompasing sociological aspect, conspiracy discourse to which I return in Chapter 8. But 'truth and lies' also plays out in the politics of experience apparent in the above examples at two levels: the substitution of scientized accounts for those of ordinary commonsense asserting the authority of science in a generalized way (Habermas 1973, see also Pollner 1987: 6-12); and the specific assertion of the superiority of scientific explanations over observers' own accounts of their experiences. However, observers may not accept the 'expert' explanation and, in Fortean discourse, it is common for resistance to the supposed authority of conventional science to be articulated. One example involving the psychology of perception concerns ouija, a type of 'talking board' (Murch 2009: 32) consisting of a display of the alphabet and a planchette used to spell out messages, often purported to be from other-worldly 'spirits'. Ouija boards have long been considered easily manipulated by users and subjected to the charge of hoaxing, but one letter-writer to FT proposed a psychological explanation, specifically 'ideomotor action' (Ashton 2009: 74). This refers to the possibility that 'merely thinking about a movement would awaken the actual movement ... below the level of conscious awareness'. It is explicitly advanced to substitute for supposed spirits, as it 'explains how ... people can be convinced of the ouija's spiritual nature ... [they] may seriously (and we could also say correctly) believe that they are not controlling their movements.' However, in the same issue, another letter on this topic also recounted a personal experience of 'table-turning' – moving a table by apparently doing no more than placing hands on top of it – and was explicit in rejecting as adequate explanations 'unconscious muscle movements, hoaxing and hysteria' (Stevenson 2009: 77). The writer was similarly explicit in doubting the authority of experts:

> The experience made me sceptical of academic experts who easily dismiss or "explain" such phenomena. I can't say what caused the table to tilt, but it would seem reasonable that the explanation lies in forces not, as yet, charted by physics. Some things can't be explained by current science, but that doesn't mean they don't happen.

Thus, within Fortean discourse both mundane self-preservation in accord with the established scientific Great Object and explicit doubt about that Great Object are to be found. Observers are often prepared to defend the validity of their experiences against contrasting accounts that seek to discount or undermine them. It is also indicative of the mixed relationship Forteans show to conventional science, both seeking scientific legitimation, but also often rejecting scientists' accounts. Sometimes science is criticized for its lack of recognition of Fortean concerns, although even where recognition seems to be granted, this can be turned into criticism as overdue and inadequate. For example, in a discussion of some recent books on UFOs, Brookesmith (2008: 49) writes: 'when academics of an anthropological or sociological disposition start to take an interest in a cultural phenomenon, one can take a reasonably safe bet that, as a cultural entity, it's all but dead.' He documents a number of instances in which arguments and conclusions reached by such 'academics' effectively repeat views expressed in specialist ufological literature without referencing these sources. Amongst his criticisms, he (2008: 51-53) refers to academics' 'ignoran[ce] of history', 'inaccurate summaries of ... cases' and lack of inclusion of non-American 'research' undertaken by 'ufological scholars'. He (2008: 50) also 'wonder[s] ... whether they were simply careful about what they *admitted* having read, lest they appear insufficiently original'. Thus, with the discussion prefigured by a reference to 'the groves of academe', it concludes that academics 'build reputations on the backs of unsung pioneers'. In this way, the superior expertize of Forteans is established and that of academics undermined. This is constructed around a moral contrast that draws on CBAs associated with scientists, broadly similar to those informing Merton's normative model (1968a), which presents an ideal-typified image of the scientific community as universalistic, communistic, disinterested and sceptically organized. Basically, these mean that scientists should assess knowledge claims not claimants, share knowledge communally, strive for objectivity, and avoid reaching conclusions until all the evidence has been reviewed. Brookesmith finds fault with 'academics' on each of these grounds: they are particularistic in their choice of referencing; socially isolated and exclusive; motivated by personal interests focused on priority and reputation; and ignorant in various ways about the evidence. Thus, he mobilizes a set of activities associated with the category 'scientist'/ 'academic' as resources with which to find their work wanting and them morally culpable, thereby turning their own rhetoric against them (see also Prelli 1989b, Taylor 1996). Forteans in contrast appear as the *real* scholars, nobly pursuing the really ground-breaking research with no concern for financial reward or social status.

Thus, whilst 'science' is often employed as a resource of legitimation, 'scientists' are often the target of criticism against which Forteans appear as more 'expert'. A similar demarcation based on relative expertize is drawn with 'the public', though here again contrasting characterizations are employed. As with the media, the public is a source of information about Fortean phenomena, but also depicted as not understanding such phenomena; accordingly they are represented as both rational and irrational. For example, in a report about a UFO conference (Pilkington 2009: 42), a prominent ufologist was quoted on the issue of government secrecy over the existence of UFOs:

> I don't think the public would have problems dealing with the reality of UFOs ... Look at the Mars microbe story from 1996. People were fascinated to know that there actually was life on other planets. If there was going to be rioting on the streets following a UFO disclosure, it wouldn't be because of the news itself, but because the government had kept the truth secret for so long.

Here 'the public' is depicted as relatively rational through a contrasting image of 'rioting on the streets'. In relation to this extreme representation (Potter 1996; see also Pomerantz 1986) of disruptive behaviour, the description of the response to the 'Mars microbe story' (concerning purported evidence of fossilized bacteria traces in a martian rock) as one of 'fascination' appears as intense curiosity, rather than, say, spellbound mesmerization. Thus, it is rationally directed and controlled. This is used to undermine the imputed motivations for supposed government secrecy about the existence of UFOs, so that it is they who appear irrational on this matter in contrast to 'the public'.

However, sometimes it is 'the public' that is presented as irrational, as in this comment by Roberts and Clarke (2009a: 28) in the same issue, regarding the cancellation of a UFO magazine described as: 'tr[ying] to present UFO related material in new and interesting ways, while keeping a raised eyebrow and slightly world-weary tone. In our opinion, it was too honest for the magazine-buying public, most of whom want pre-digested glossy belief systems given to them on a plate.' Here, an Adornoesque (Adorno 1991) view of the public is given, albeit modalized (Latour and Woolgar 1986) as an 'opinion', contrastingly formulated against an image ascribed to the magazine that invokes a sense of critical distance from UFO material. So, as in the standard terms of mass culture theory discussed in Chapter 1, the public appears as infantile and unreflective, mesmerized by the 'glossy' superficialities peddled by the mass media. Thus, far from being rational and self-controlled, now they are irrational victims gulled by a 'dishonest' consumer culture.

Thus, Fortean discourse parallels the discourse of scientists in constructing boundary demarcations between Forteans and 'non-experts', which can include both scientists and the public. This enables a politics of experience in which Fortean accounts can be justified as superior representations of reality, on both the grounds that they are supported and not supported by either scientists or the public. Both categories are open to mixed attributions that can in either case be used to preserve the Fortean version of reality. If scientists agree with the Fortean version, then this shows its superiority is vindicated at last by the Johnny-come-latelies; if they disagree, then this shows their ignorance. Similarly, if the public agree with the Fortean version, then this supports the fringe against the powerful orthodoxy, while if they disagree, then this is no more than to be expected from the gullibly ignorant.

However, these external divisions are accompanied by internal divisions that further refine these distinctions. Here, Forteans work to position themselves in an intermediary territory between the categories 'pseudo-scientist' and 'debunker' (see Chapter 6). Thus, while Forteans may align themselves with science as a generalized category (Parker 2001: 196) appearing sceptical towards fringe science claims – or in some cases, such as 'intelligent design' (Simmons 2006), rejecting them outright – they are also concerned to distinguish themselves from two other groupings: 'believers', and 'skeptic[s] with a "k"' (FT204, December 2005: 1). This latter category is especially associated with the Committee for Skeptical Inquiry (CSI)[1] – hence the 'k' – and for Forteans is synonymous with 'debunking'. Thus, Forteans work to construct a third position that is fully aligned with neither.

In part, this is done using the standard tropes of disenchanted mundane reason, as can be seen in an article by Paul Devereux (2007) reviewing the history of 'leylines'. The term 'leys' was originally coined by Alfred Watkins (1974[1925]) to refer to what he identified as straight-line connections between sites of ancient or sacred monuments and other prominent landscape features, sometimes stretching for many miles over the British countryside. Watkins speculated that these were old travellers' ways, but Michell (1973, 1974) linked them with the then popular idea of 'ancient astronauts' – alien visitors to earth in ancient times (Pauwels and Bergier 2001; also von Daniken 1971) – to argue that leys were lines of energy running through the surface of the earth defining a 'sacred geometry'. These ideas caused some controversy with orthodox scientists (for example, Krupp 1984) and Devereux (2007: 32) refers to 'heated debates with archaeologists and statisticians about the relationship between claimed leys and chance alignments', the outcome of which was that 'energy ideas are shown as being supported by claims rather than accountable research'. Thus, in addition to orthodox scientists, he (2007: 33) distinguishes 'a three-way split ... between ... traditional Watkins-style ley hunters, ... critical, research-based ley hunters, and ... dowsers and New Age pulpiteers popularising the notion of ... energy lines'. This was 'exacerbated' when he himself 'adopt[ed] an anthropological approach' that 'explain[ed] the untenable nature of "energy line" hypotheses' and brought in the notion of 'spirit roads' derived from traditional folklore and shamanistic practices.

Within all this, then, a number of different categories are distinguished both between Forteans and non-Forteans, and amongst Forteans themselves. To construct these divisions, Devereux draws on the rhetoric of science to position himself within the order of scientific disciplines, separated from 'archaeologists and statisticians', but aligned with 'anthopology' through a self-description as 'research-led'. Thus, he (2007: 30) opens the article with a contrast between 'Watkins's ... vision' and what 'has become increasingly obvious to research-minded ley students', to wit that 'leylines' do not exist. This conclusion he characterizes as 'follow[ing] where the mythical leys led', echoing a standardized verbal formulation employed by

1 CSI was founded in 1976 as the more self-descriptive Committee for Scientific Investigation of Claims of the Paranormal (CSICOP). Its website states: 'The mission of the Committee for Skeptical Inquiry is to promote scientific inquiry, critical investigation, and the use of reason in examining controversial and extraordinary claims.' (CSI 2010)

scientists (Gilbert and Mulkay 1984: 64). Thus, Devereux uses the rhetoric of empiricism to forge distinctions amongst scientists within which to position himself. In addition, he uses the rhetoric of error accounting in the form of critical rationalism to construct divisions amongst Forteans, specifically to demarcate those who are 'critical' in their 'hunting' for 'leys' from both those who 'hunt' in a 'traditional' manner and those who 'pulpiteer'. Thus, the lines are drawn (no pun!) exactly in the style of the disenchanted mundane discourse of modern science, marking distinctions from both received wisdom and religious dogma on the grounds of evidential and rational appeal. This shows that this discourse, rather than simply solving 'mysteries' by making the unknown known, enables ever finer hairs to be split, fraying and complicating the fringe (see also Nixon 1999).

However, whilst Forteans often mark their distinction from 'believers' like ley-hunters, they also distinguish themselves from debunking 'skeptics' – or 'pseudo-skeptics' as they were called by the sociologist Marcello Truzzi (2006: 59), a founding member of CSICOP who resigned finding it too 'absolutist'. This is a recurring feature of Fortean discourse, as in this example by one of its founding editors, Paul Sieveking (2006: 73):

> we are all believers in one way or another – in one thing or another – although ... immersion in forteana tends to loosen fixed ideas in favour of temporary acceptance ... "pseudo-skeptics", by contrast, are particularly stubborn believers in materialist reductionism and argument from authority, and hate their prejudices to be disturbed by awkward data.

Again, there is here an empiricist appeal in the reference to 'awkward data', which is couched against a commonplace formulation preserving mundane idealizations. The expression, 'we are all believers ... in one thing or another' is used to resist the possibility of a condition of radical doubt about the existence of a world that is determinate and knowable in principle. The exteriority of a world of 'things', of 'data' is an incorrigible proposition that defines the limit of doubting and sets the terms within which this can occur. Thus, although we may doubt some specific things, like the claims of 'believers', we cannot doubt the existence of all things because we must still be believing in something. The puzzle of unknowability, of 'mystery' is therefore resolved by shifting determinateness from the outside world to inside our heads. Even if the externality of specific things cannot be retained – and thus mystery persists – the internality of belief in something can – and thus conviction about the in principle knowability of reality be upheld. The mundane idealization in determinateness is preserved despite reality remaining unknown. The necessity of this is accentuated by the extreme case formulation, 'all' (Pomerantz 1986), which makes it an inescapable mental condition and thereby normalizes it (Wooffitt 1992) – and so echoes Weber in its definition of the condition of the ordinary person in modernity. Thus, because we have to be believers in things, there is no error in being so or accepting this.

Where there is error, however, is in being 'stubborn' about our beliefs, slavish disenchanters doubting neither 'authority' nor 'materialism' – the characteristics of 'skeptics', who as such are not really sceptical at all. In contrast, Forteans allow belief only 'temporary acceptance', not letting their 'prejudices' cloud the appearance of 'awkward data'. So, despite the incorrigibility of the proposition that things exist, Forteans work to open up discursive space in which the nature of those things remains mysterious – they are at best only temporarily known. Thus, for all its adoption of the mode of reasoning associated with disenchanted mundaneity, Fortean discourse also allows questioning of disenchanted mundane idealizations. The exemplar for this is Fort himself, as represented for example in an article by Wilson (2008: 57). Like Sieveking, Wilson draws a sharp boundary-line between Forteans and the 'arch-sceptics (or skeptics)' of CSI, who are 'anti-fortean': 'The Universe described by the sceptics is highly ordered and organised, and has no place for the eccentric and incongruous, which explains why sceptics see Fort and his readers as a threat to the natural order.' The crucial difference for Wilson is the form of 'humour' adopted: 'sceptics' use 'ridicule' and 'mocking laughter', where Forteans are 'merry-making'. Accordingly, where 'sceptics' employ mockery to 're-establish order' (see also Billig 2005) asserting '[r]ight is right and wrong is wrong', Forteans must 'encompass ... both worlds at once ... we have to treat them as real and unreal, rational and irrational, acceptable and unacceptable at the same time ... Forteans should be in equal measure gullible and sceptical' (Wilson 2008: 56). The model for this is Fort himself, whose 'whole oeuvre has the structure of a joke', one in which he 'includes himself and his explanations.'

There are intriguing parallels here with the Strong Programme in SSK (Bloor 1976: 4-5) – specifically the tenets of impartiality, symmetry and reflexivity – and a view of Fort as having 'prefigured' such ideas has been noted in the pages of FT (Kidd 2006: 55). These should not be pressed, however, for although Fortean discourse remains deeply informed by the rhetoric of disenchanted mundane reason that insists upon the in principle knowability of the world as a determinate, coherent and self-consistent Great Object, it also strives to hold open discursive space for 'mystery'. In the words of Patrick Harpur (2009: 60-61) describing his approach to forteana: 'I did my best to avoid "explanationism" ... A mystery isn't a problem. It can't be solved – it can only be entered into. And you are changed as a result.' So, whilst we may be obliged to believe, it may not always be the same person doing the believing and perhaps then there is no more determinateness to be found inside our heads than out.

Chapter 8
Rationalized Blaming: Conspiracy Discourse

The last chapter looked at aspects of fringe science through the lens of Fortean discourse to argue that, despite its orientation towards the in principle knowability of reality, disenchantment actually engenders 'mysteries', a discursive space constituted as 'the unknown' within which a form of non-disenchanted and non-mundane reasoning persists in modern culture. A key role is played in this by the types of mundane self-preservation that characterize the disenchanted discourse of science, centred around accounts of error in the form of technologies of detection, psychologies of perception and sociologies of conception. These are designed to preserve the incorrigible propositions at the core of disenchaned mundane reasoning – that the world is determinate, coherent and non-contradictory – in accord with what is taken to be the prevailing Great Object of orthodox science. Defined negatively, this Great Object excludes the possibility of 'other-worldly' phenomena or entities and much fringe science is marginalized on the grounds that it might impute the existence of such things. However, accounts of error are taken up by both sides, enabling controversies to persist and grow in their speculative compass to account for sustained denial in the face of what is taken to be evident and rational. One outcome of this is a particularly wide-reaching form of the sociology of conception that has a niche position within Fortean discourse – conspiracy discourse.

A number of writers have argued that conspiracy discourse has become more widespread in contemporary culture (Dean 1998, Knight 2000, Parish and Parker 2001, Barkun 2003, Latour 2004, see also Locke 2009c) and one view is that this expresses anxieties generated by the paradoxical consequences of instrumentally-rationalized modernization self-undermining through the creation of instability and risk (Parish 2001). However, it has also been commented that the attempt to account for conspiracy discourse by reference to powerful, hidden social forces looks more than a little like a form of conspiracy theory itself (Parker 2001). Conspiracy theories, then, present professional sociology with a particular discomfiture, as not only are they forms of 'social and cultural theory' (Bell and Bennion-Nixon 2001: 149) that threaten its claim to a monopoly over the right to produce legitimate knowledge of the social, but their similar structure of argument (Latour 2004: 229) resists incorporation into established sociological Great Object(s) by the simple mechanism of ironizing the ironizer (Knight 2000). As such, what conspiracy discourse confronts sociologists with is the problematic nature of our own adoption of mundane idealizations, posing us the question of how to account for the unaccountable without discounting our capacity to provide an account. The answer is by constructing an account that accounts for itself.

Now, this is exactly what Weber sought to do as a response to the problem of the sociology of knowledge through the thesis of rationalization. However, as I have argued, this amounts to a moral plea about the way scientists should conduct themselves that downplays significant features of their discourse, specifically enchanted rhetorics and accounts of error. Further, both the rhetoric of disenchantment and the uptake of accounts of error in public contexts contribute to conspiracy discourse. Nonetheless, I want to suggest that in this resides the possibility of an alternative sociological understanding that can account for both this and its own development. This requires recognition of the moral character of sociological reasoning, not as Weber argued as an ethic of conduct that strives to avoid its own ethical constitution, but as one that recognizes itself as formed from and within accounts of suffering. This raises the question of what form accounts of suffering take under conditions given by disenchanted mundane reasoning, a question that can be approached through considering ways of attributing blame.

Conspiracy Discourse and Social Theory

In accord with their appeals to 'scepticism', Forteans often express doubt about the claims of conspiracy theorists, but conspiracy discourse nonetheless has a place within Fortean discourse for two main reasons: first, because Fort's own views often ran in directions that resonate with conspiracy theory (Hierophant's Apprentice 2007) and second, because some conspiracy discourse grows out of fringe science (Barkun 2003: 27). An example of the latter is ufology, in which official denials of the existence of any unexplained phenomena seemed to conflict with official interest in them (Dean 1998, Denzler 2003), while official accounts themselves have sometimes appeared contradictory. One of the earliest and most famous cases illustrates this. In Roswell, New Mexico in 1947, what was first publicized as an official story of the discovery of a flying 'disc' was rapidly re-reported as officially being 'nothing more than an ordinary weather balloon' (Evans and Stacy 1997: 6; see also Denzler 2003). Such apparently contradictory accounts ostensibly conflict with mundane idealizations, but could be explained as due to errors such as insufficient knowledge of weather balloons, misunderstanding of intended meaning or misreporting. However, they can also be seen as an attempt to cover-up an unintentional leak or as intentional disinformation designed to keep the public in a state of uncertainty, accounts that move in the direction toward conspiracy. According to Evans and Stacy (1997), the weather balloon story was not widely questioned until the early 1980s, when a conflicting report attributed to a military eye-witness was publicized. The intervening period, however, had seen conspiracy discourse become an established feature of ufology, because of persistent official denials of what seemed a widespread public experience. Dean (1998: 34-46) argues that at the core of this was a politics of experience in which scientists, acting as debunkers, struggled to assert their authoritative position as legitimized definers of reality against members of the public asserting their right to be taken seriously as citizens. This resulted in growing distrust of both scientists and the government amongst the relatively small community of UFO believers. However, conspiracy discourse was also being fed from other areas of fringe science, such as mind control involving experimental drugs and technologies, and these combined with themes drawn from 'radical right' political groups concerning 'secret societies' like the Illuminati and the New World Order to produce an increasingly diverse and complex mix of worldviews (Barkun 2003: 75-87).

Thus, fringe science contributes to conspiracy discourse and to an extent this can be understood as simply an extension of utilizing sociological accounts of error. Fort (1998: 136) himself provides some illustration of this in that his distrust of established science and other institutionalized authorities led him to give an account of their collective failure that sounds a little like a pre-echo of Kuhn (1970): 'I conceive of nothing in religion, science, or philosophy, that is more than the proper thing to wear, for a while'. In more technical sociological terms, we might prefer to speak of socialization into a dominant ideology (Abercrombie, Hill and Turner 1980), but the basic postulate of a single, authoritative group controlling ideas and systematically hiding the truth to protect its own interests is identical.

The parallels can be brought out further using Barkun's (2003: 3) definition of a 'conspiracy belief', as 'the belief that an organization made up of individuals or groups was or is acting covertly to achieve some malevolent end'. Underlying this is an attempt 'to delineate and explain evil' and a view of the world as 'governed by design rather than randomness' (see also Byford 2002: 3.3). This is apparent in three common 'principles' of conspiracy beliefs: 'nothing happens by accident'; 'nothing is as it seems'; and 'everything is connected' (Barkun 2002: 4). Thus, conspiracy beliefs tend towards totalistic explanations that incorporate all social events and activities; or, as one writer put it (Kossey, cited by Hierophant's Apprentice 2007: 52): 'Conspiracy theories are like black holes – they suck in everything that comes their way .... Everything you've ever known or experienced, no matter how "meaningless", once it contacts the conspiratorial universe, is enveloped by and cloaked in sinister significance.' Or, in Spark's (2001: 48-49) more prosaic description, conspiracy theories have an 'excess logic' marked by 'a profligate mish-mash of detail'.

For Barkun (2003: 6-7), this feature is particularly characteristic of what he calls 'superconspiracy theories' and is a source of both strength and weakness: its strength is that it accords with the 'scientific' principle of 'parsimony' by reducing the complexity of social life to 'a single plot'; but its weakness is that this very reduction 'defeat[s] any attempt at testing'. The conspiracy is at work behind all sources of information and all systems of knowledge production, therefore despite appealing to empirical veracity through 'elaborate presentations of evidence', anything that may be advanced as disproof can be discounted as part of the conspiracy designed to distract attention. Even official evidence confirming some aspect of the conspiracy may be discounted as 'disinformation' designed to draw attention away from the *real* conspiracy happening elsewhere. Thus, the theory is 'unfalsifiable' as both ostensible proof and disproof can be interpreted as part of the conspiracy. The obvious question that then follows is how the believers themselves have managed to disclose the plot. Two major strategies here are to claim insider knowledge of the conspirators' activities, or conversely to maintain outsider status and so freedom from the 'mind control and brainwashing' affecting everyone else. It is this latter feature that facilitates the link to what Barkun (2003: 8) calls 'stigmatized knowledge' such as fringe science.

The similarities between these features and those of 'critical' social theory are readily apparent (Latour 2004). Essentially, a 'superconspiracy' is simply a type of 'grand theory' (Skinner 1990) that postulates the lurking presence of a pervasive set of interests informing all aspects and levels of society, typically associated with one specific group such as capital, patriarchy or the neo-colonial race. Under such a dominant power, 'nothing happens by accident', because the group asserts its control 'three-dimensionally' (Lukes 2005) to minimize the possibility of social action outside its decision-making. Thus, 'everything is connected' systemically through social networks that interconnect the members of the ruling group in a web of unseen control (Scott 1991). Further, 'nothing is as it seems', because the group maintains its dominance through control over sources of knowledge and means of information which are 'systematically distorted' (Habermas 1971) in its favour. These include science – which is shaped by capital (Aronowitz 1988) and/or patriarchy (Harding 1991) and/or post-colonialism (Amsler 2007) – education (Bowles and Gintis 1976), and the mass media and popular culture (Garnham 1997). However, its invisible influence is yet more insidious as it asserts control even over the language we speak, setting structural limits to our capacity to raise and test validity claims (Habermas 1984, 1987), and defining the 'order of things' within which power/knowledge is constrained (Foucault 1970). Our fundamental 'categories of thought' (Beck 1992) are shaped by these encompassing interests, so raising the question of how some sociologists have managed to rid themselves of their influence sufficiently to see through them. Like conspiracy theorists, sociologists' strategies to account for the privileged status of their knowledge include, claiming 'insider' information generated by special research techniques such as participant observation (for example, Collins 1983), or conversely claiming 'outsider' status as 'strangers' unaffected by the taken for granted assumptions of the dominant group (for example, Latour and Woolgar 1986). Similarly, sociologists often provide extensive empirical documentation to support their claims, subjecting this to meticulous scrutiny to disclose the heretofore unrecognized presence of the hidden interests at work in the background. However, for critical sociologists, the notion of 'testing' knowledge claims is inappropriate as this itself is constituted within the terms of the dominant ideology or discourse. 'Testing' knowledge presumes the existence of a world of 'facts' independent of social constitution and observers capable of being 'objective', neither of which is possible as both the phenomena of the social world and our capacity to make sense of them are shaped by dominant interests.

The parallels reach further, however, when we consider how sociologists account for the apparent growth and popularity of conspiracy theories. Traditionally, conspiracy discourse has been linked to 'paranoia', the term originally used by Hofstadter (1965) to account for 'radical right' political groups in the United States (Barkun 2003) and still employed by more recent writers such as Spark (2001: 49; see also Pipes 1997, Dean 1998). Hofstadter did not intend the term to be psychologically reductionist, but to describe a collective reaction to a situation of social instability affecting marginalized groups, who perceived their established way of life to be under threat from powerful outside forces (Knight 2000). Nonetheless, it has been taken up as an invitation to provide psychological explanations predicated on the assumption that conspiracy discourse is essentially erroneous, if not pathological (for example, Kruglanski 1987, Clarke 2002; see also Byford 2002). Knight (2000) argues that both error and pathology are implicit in the term and indicative of the assumption that conspiracy theories can only arise from some sort of aberrant condition generating an 'irrational' response. Moreover, it implies a condition of 'normality' that is effectively equated with liberal pluralism, which whilst it may be sympathetic to the 'errors' of minorities, has little tolerance for the 'extremes' of white supremacists or anti-Zionists, who in Pigden's (1995: 6) view are simply 'demented'. Thus, describing conspiracy discourse as 'extreme' often conflates two senses, referring to both its 'pathological' thinking or 'excess logic' and its right-wing politics; that is, to reason and morality which are effectively aligned. So, in ascribing conspiracy discourse to 'paranoia', a base-line norm of both acceptable behaviour and acceptable thinking is implicated.

In short, this is a classic exercise in professional boundary maintenance (Gieryn 1983, 1999), involving a politics of experience (Pollner 1987) in which the superiority of the analysts' account of reality is asserted over that of 'lay' members. This applies also to sociological accounts that see the growth of conspiracy discourse as a response to the 'postmodern' condition of social fragmentation (Dean 1998) or the 'anxieties' of 'risk' society (Parish 2001). Such accounts differ from Hofstadter's in that they do not see conspiracy discourse as socially marginal, but as a widespread consequence of the absence of 'consensus reality' (Dean 1998: 8) brought about by social fragmentation. However, whilst this might account both for the growing diversity of conspiracy discourse – which is no longer, if ever it was, confined to the 'radical' right (Knight 2000, Sparks 2001) – and its attraction towards 'stigmatized' knowledge (Barkun 2003: 26-29), it makes it difficult to understand the all-encompassing scope of 'superconspiracies' and their appeal to the standards of formal scientific discourse. Why would a credible response to social fragmentation be an order of discourse constructed around absolute social singularity? And why do conspiracy theorists cling to the canons of empiricism and rationalism associated with established, modernist science? After all, Lyotard (1984) forecast that, in the postmodern condition, people would embrace the general agonistics of the multiplicitous language-games brought on by the collapse of the grand metanarratives of science (see Chapter 2), but conspiracy theorists seem to do the opposite. The implication again, then, is that conspiracy theories are both erroneous, because they do not correctly understand their social world believing it monolithic when it is polymorphic, and abnormal, because the 'norm' should be to discourse in flexible, multivarious and hybrid ways. Thus, seeing conspiracy discourse as a response to postmodern fragmentation is not so different from seeing it as a marginalized 'paranoid' reaction to social instability. A similar point applies to seeing it as a response to the anxieties of risk society, which has clearer parallels with the 'paranoid' view, the only difference being that the 'paranoia' is seen as entirely justified! The risks are real and the anxiety therefore understandable; nonetheless, conspiracy discourse is tilting at the wrong target(s) as the sociological analyst is able to show. Thus, conspiracy discourse is still erroneous and remains pathological, even if it is symptomatic of a more general social illness.

These accounts, then, do not escape the politics of experience; rather, they continue to assert the greater validity of the analyst's account, treating conspiracy discourse as essentially erroneous and symptomatic of moral-normative breakdown. Part of the problem with this, however, is that conspiracy discourse is difficult if not impossible to disprove, since any evidence that might be offered against it can simply be discounted as part of the conspiracy. Moreover, conspiracy theorists can point to evidence in support of their claims and they argue in a fashion that even Hofstadter described as 'rationalistic' (cited by Barkun 2003: 29; see also Groh 1987: 4). Similarly, Byford (2002: 3.4) states that at least some types of conspiracy theory 'simulate mainstream analyses', but this begs the question of how we distinguish a simulation from the real thing. If it is the case, as Knight (2000: 16) asserts, that '[a]ctual conspiracies of some stripe have undoubtedly existed at various times', then it is not so easy to begin from the premise that conspiracy discourse is simply erroneous. Instead, we might try like Pigden (1995) to distinguish the bits that are valid from the bits that are not – but then we return to the first problem of trying to refute the irrefutable in the light of which any selection we make appears arbitrary. Further, the problem is compounded by the manner in which conspiracy theorists 'simulate' 'rational' analysis. Knight (2000: 16-17) puts this rather differently, pointing out that conspiracy theorists have appropriated the 'diagnostic language' of scholars and used it to turn the tables on their critics by ironizing the ironizers: 'The increasingly self-reflexive "paranoid" narratives have … begun to internalise the modes of reading traditionally brought to bear upon them, to anticipate and disarm the authority of expert criticism'. Thus, they question whose interests are being served by the 'expert' accounts linking them to the conspiracy: 'What better way … to divert attention from … [the conspirators] than a well-respected academic telling everyone that none of it is true, and, what's more, that you're a touch hysterical if you believe it?'

Conspiracy Discourse and/as Mundane Reasoning

Thus, any attempt by sociologists to account for conspiracy discourse by reference to hidden, unseen social forces can simply be turned around, because for the conspiracy believer it is the reverse that holds. This is exactly the situation described by Pollner (1987: 77) regarding the mutual undermining of conflicting experiential accounts: 'each of the disputants, treating [their] version as a given and thereby ironicizing competing experiences, finds the experiential claims of the other to be the product of an inadequate procedure for perceiving the world.' Like critical sociologists, conspiracy theorists view the world through mundane idealizations as determinate, coherent and non-contradictory, and use techniques of self-preservation to maintain whatever incorrigible proposition they are built around. Thus, for example, the claim that the United States government has secret knowledge of a crashed UFO from Roswell can be upheld despite any kind of evidence that might contradict it, whether in the form of government denials (well they would say that wouldn't they?), the release of official documents that show no evidence of a cover-up (how do we know what is not being released?), or even official admission that secret research into UFO reports did take place (disinformation) (Evans and Stacy 1997: 7-8; see also Denzler 2003). Further, the 'black hole' like tendency of conspiracy theories to absorb everything and grow in scope is characteristic of the mundane idealization of the Great Object. As discussed in Chapter 7, mundane reasoning treats objects as self-consistent both in themselves and in relation to their context, but this includes a potentially infinite range of features – anything that happened or appeared in the situation. Thus, everything becomes potentially relevant, imbued with meaningfulness in relation to the imputed explanatory schema. To describe this as a 'profligate mish-mash' discounts the reasoning involved, which seeks self-consistency at every level from the original object, to the 'object-in-context' and on. It follows that once any one feature of the situation is called into question, the way is potentially open for every other feature to be re-interpreted in terms of the conspiracy view.

Some idea of how this works can be drawn out from Sacks's (1995a: 42) comments regarding the assassination of President Kennedy. The immediate concern, he argued, was to identify the 'category [of member] … that performed the act', that is who did it, not as a particular person, but crucially as a relevant category. Not just anyone shoots Presidents; indeed, not even just any shooter shoots Presidents, but only a special category of shooter – an assassin. Accordingly, the person responsible needs to have the relevant qualities associated with this category; they need to be identifiable as being a possible and likely assassin. Thus, once Lee Harvey Oswald was singled out, the category was crucial in deciding the matter: 'the basic character of the argument … turns on the issue of some people saying, "Of course if he's a Communist he would have done it," and others saying, "The last person who could have done it would be such a one and therefore he couldn't have done it".' (Sacks 1995a: 180) In other words, did Oswald have the relevant qualities to be this particular assassin? One such possible quality was being a Communist and, in the eyes of some United States citizens at the time, this may have been a sufficient qualifying feature. For others, however, it was not, or Oswald may have been viewed through a different category with qualities neither associated nor typically associable with such an assassin. To satisfy mundane idealizations, the object-in-context, 'assassinated-President-with-an-assassin' has to be coherent and non-contradictory, so the assassin has to display qualities consistent with an opponent of a President, such as an opposing political stance. If, however, this is disputable, then the reality disjuncture created would open up the potential for questioning not only the object-in-context, but also the account of it given. Thus, questions might begin to extend beyond merely asking whether Oswald really fit the category, towards asking why this is the official version and what then are we not being told. The conspiracy then has begun.

Thus, as Pollner (1987: 71) mentions, conspiracy is one resource of mundane self-preservation. This has quite different implications for how we think about it sociologically to the view that it arises from 'paranoia' or 'anxiety'. Seeing it in mundane terms shows conspiracy discourse to have nothing abnormal nor pathological about it in itself – it only becomes so when ironized by official versions asserting their privileged status as definitive accounts of reality. But this leads to the potentially endless cycle of mutual discounting and self-affirmation, which does not help us understand why conspiracy discourse persists. Viewing it through the lens of disenchanted mundane reasoning, however, opens up a different possibility, by taking us back to the root of Weber's understanding of rationalization in accounts of suffering.

Conspiracy and Blaming

As seen above, Barkun explicitly defines conspiracy beliefs as attempts to 'explain evil'; in Weber's terms as seen in Chapter 2, they can then be seen as a type of theodicy accounting for suffering. A similar view comes from Groh (1987: 1):

> Human beings are continually getting into situations wherein they can no longer understand the world around them. Something happens to them that they feel they did not deserve. Their suffering is described as an injustice, a wrong, an evil, bad luck, a catastrophe ... In the search for a reason … they soon come upon … an opponent group to which they then attribute certain characteristics: This group obviously causes them to suffer by effecting dark, evil, and secretly worked out plans against them.

Groh (1987: 11-12) goes on to express puzzlement that the 'cognitive reorientation', which produced the 'universal-historical caesura' associated with western rationalization has not confuted this kind of 'irrational' thinking. The reason I suggest is because rationalization has not curtailed the activity of accounting for suffering, but simply redirected it and conspiracy discourse is one form it takes under conditions set by disenchanted mundane reason.

A way into this is through the notion of 'blame culture'. Over recent years, this term has entered public discourse in Britain, often to describe organizations said to be characterized by mutual distrust and blaming in their internal working relations. Such organizations are said to have a 'blame culture' in the sense of a *culture of blaming*. However, this sense of the term might be taken as one example of a wider phenomenon that includes other types of 'culture', such as 'compensation culture', 'gun culture' and so on. In this sense, 'blame culture' refers to the *blaming of culture*. This also seems to be a feature of recent public discourse, pointing to some form of 'culture' as responsible for perceived social problems of various kinds. However, there is a significant difference of emphasis between the two. A culture of blaming implies a tendency to seek a specific target to which responsibility can be assigned, as is apparent where compensation is being sought in that some*one*, whether an individual person or an organization, needs to be identified as the focus of blame. The action of specified individual(s) is held responsible and concomitantly that or those individuals are assumed to have responsibility for their actions. Moreover, the actions in question are likely to be relatively well-bounded, to be specific actions in specific circumstances with specifiable consequences. In the blaming of culture, however, just the opposite is the case as this effectively avoids blaming specific individual(s), who are instead seen as representatives of something greater. It is this general 'thing' that is held responsible and individuals are subject to this greater force that holds them in its sway. Accordingly, a range of individual actions across a variety of situations and circumstances would be candidates for inclusion under the general category being blamed.

These contrasting modes of blaming emerge out of a fundamental dilemma informing modern culture regarding the attribution of responsibility (Yearley 1985), which provides a resource for constructing accounts of suffering. This can be understood in Weberian terms as a consequence of disenchanted rationalization, at least once this is re-crafted in the manner outlined earlier. As discussed in Chapter 2, Weber's analysis of 'intellectualist rationalization' develops from his philosophical anthropology (Poggi 2006) that sees human beings as confronted by the existential need to account for suffering, which led to the development of religious beliefs as the unfolding discourse of the charismatic in 'this-worldly' and 'other-worldly' directions. In the west, this-worldly rationalization has culminated in the modern period in a disenchanted mental orientation that displaces all other-worldliness with a relentless belief in the in principle knowability of the world manifest most fully in modern science. The paradox of consequences in this is that a form of value-rational action and expression of the charismatic has given rise to a type of mental outlook that displaces matters of moral principle with the valorization of pure instrumentalism having no regard for ultimate values or belief in sacred cows. However, as discussed in Chapter 3, this analysis contains an ambiguity. Given that *Zweckrationalität* is nonetheless a form of *Wertrationalität*, then the relation between the two need not necessarily move in only one direction.

Weber tended to accentuate the erosion of *Wert* by *Zweck* because of his concern over the ethic of conduct of the scientist, but Whitehead (1974) brings out the counter-side to argue that we should also expect to see a process of the continuing abstraction and universalization of the charismatic. In other words, rather than assuming the fundamental urge to account for suffering has been eroded by a general condition of disenchantment, we should instead think about what character accounts of suffering might take under circumstances where the charismatic has attained a heightened level of abstracted universalism. Further, we need to think about this in relation to the emergence of a specifically disenchanted mode of understanding as expressed especially in the discourse of modern science.

Here we can begin to see how blame culture might be understood as an outcome of these developments. Accounts of suffering are, at one level, ways of attributing blame, providing solutions to the basic question of why people experience this-worldly difficulties – pain, oppression or trying circumstances of one kind or another. Historically, there are three broad loci towards which blame has been directed in different times and places: god, nature and people.[1] In the modern outlook of disenchantment as Weber sees it, god is no longer available as a locus of blame (a similar view was taken by Popper as cited by Pigden [1995: 7-8], though it should be noted that some contemporary conspiracy theories do blame the Antichrist [Barkun 2003: 39-64]). Rather, it is only an impersonal nature to which recourse may be made. However, as Weber's (1948b) lecture on the vocation of science acknowledges, the very impersonality of nature makes it a difficult choice to accept, as essentially it means giving up the capacity to assign blame at all: suffering is merely the outcome of a chance concatenation of circumstances, a probabilistic distribution – there is nobody and nothing to blame, just bad luck in a universe devoid of meaning. Accordingly, there is no means of redress, no ethic of conduct or path to salvation. Such 'salvation' as there is resides in duty, but it is a hollow performance, a duty to nothing except its own undertaking. This is the attitude of mind that unites the scientist and the bureaucrat in a spiritless vacuum of disciplined routine.

However, as already suggested, there are grounds for thinking otherwise since attributions of blame continue in at least one form – accounts of error or mundane self-preservations. As argued in Chapter 3, accounts of error in scientists' discourse refer to cognitive or methodological errors, and personal and sociological factors roughly corresponding to Weber's definitions of 'formal irrationality', 'substantive irrationality' and 'substantive rationality' respectively. Notably, the latter two include a normative dimension involving what Weber (1968: 656) called 'ethical imperatives'. Meanwhile, Pollner (1987: 62) argues that mundane self-preservations may be directed at the level of the object, the experience or the account. Broadly, then, they may be ontological (are the objects real or the same?), epistemological (is the observer's knowledge claim credible given their capacity

1 Douglas (1992) suggests an alternative three based on anthropological studies: me, you, or them. In the modern world, however, we should also include 'it'.

to experience or comprehend reality properly?), or moral (is the observer lying or otherwise deliberately misrepresenting things, perhaps for reasons of personal bias or group commitment?). Looked at from either view, scientists' discourse – which is also taken up by wider publics – and ordinary mundane reasoning both contain available resources of a moral nature, not merely to account for error, but also to attribute blame on the basis of right or correct conduct.

Thus, the capacity and means to assign moral responsibility is not eroded by disenchanted mundane reasoning, merely redirected in its locus of blaming. Within the rules of the language-game of this discourse, blaming god is, if not strictly impossible, then at least an incomplete move. Meanwhile, blaming nature, whilst the preferred move, is ultimately unsatisfactory because a disenchanted nature is empty of motivation and purpose; it is amoral and therefore not really *to* blame. That leaves people. Since the Enlightenment at least, however, people have been open to a dual understanding as both individual and collective (Durkheim 1964). This, then, generates a dilemma in the attribution of blame (Billig et al. 1988): do we blame an individual or a collectivity? Do we engage in a culture of blaming, or a blaming of culture?

Suffering Sciences

Thus, there are two cross-cutting axial dilemmas for the disenchanted mundane reasoner: one between locating blame in nature or in people; and one between attributing blame to an individual or to a collectivity. This produces a four-fold framework of tendencies of moral reasoning from which resources for attributing blame may be constructed. Some support for this comes from the human sciences themselves, treated as extensions of the mundane concern with attributing moral accountability within the terms of the logic of disenchantment. These sciences are themselves developments out of this logic, articulations of the discourse of blame attribution in modernity conforming ideal-typically to the four-fold framework of tendencies of moral reasoning: nature + individual ('behaviouristic'); nature + collectivity ('systemic'); people + individual ('therapeutic'); people + collectivity ('conspiratorial').

Behaviouristic

At one end of the axis of moral accounting is nature, but since disenchanted nature is amoral there can in a sense be no blaming. However, because disenchantment is still a form of value-rationality, albeit a peculiar one in which values are depersonalized and instrumentalized, individuals can still be called to moral account on the grounds of being insufficiently or improperly calculative. This is seen in science in respect of methodological errors (see Chapter 3), but in application to individual action in general, the meaning of 'moral' action is simply to act in a purely instrumental, calculative way like a behaviouristic machine.

This is ideal-typified in the rational economic actor in whom morality is reduced to a pure cost-benefit calculation: the good actor is the economic maximizer – and this is the type of moral reasoning informing economics in its neo-classical 'paradigm' (Swedberg, Himmelstrand and Brulin 1987, Heelas 1991). Thus, ideal-typically, the individual behaves like a computer following a programme and they can no more be blamed for their actions than a machine. Rather, blame may be attributed for *not* behaving sufficiently calculatively or instrumentally, specifically for allowing 'subjective' values and moral principles to interfere with functional procedures. This is not so much a moral judgement, however, as a calculative one, penalized by failure in the economic competition of all against all. Salvation then resides in doing your sums better in future.

Systemic

On the other hand, where the spiritless spirit of disenchanted morality is informed by a logic of collectivism, a different mode of moral accounting develops. Here, morality again takes a reduced, depersonalized form, but because it is applied collectively, rather than individuals behaving instrumentally, it is an instrumental logic that impels individual behaviour. In effect, individuals become bearers of an instrumental reasoning beyond their personal intent, but which nonetheless shapes what they do. The ideal-type of this is sociobiology (Wilson 2000; see also Lyne and Howe 1990), in which individual action is reduced to a mechanistic outcome of processes directed by and towards genetic, or genomic self-reproduction. Here, moral accountability resides not with individuals, but the genetic survival machines impersonally – if 'selfishly' (Dawkins 2006) – directing their host bodies. Again, there is no morality in the traditional sense here, simply a generalized logic of encoded machinations endlessly re-forging chemical building blocks out of chance reactions – a sort of *Legoland* morality. Accordingly, no blame as such is to be attributed; 'blame' equates to unfit mutations and these are simply eradicated by the impersonal expedient of natural selection (Popper 1972). However, there is a sort of 'salvation' to be found in assisting the process of genetic selection to eradicate 'unfit' modes of behaviour (or even whole subspecies) by engineering them out of the gene pool or modifying them with appropriate chemical nostrums (Nelkin and Lindee 1995).

Therapeutic

At the other end of the axis of moral accounting are people. Here, in contrast to the amoral reductionism of disenchanted nature, at the opposite extreme morality informs all action and everything is a sign invested with moral import. People are thought of less as extensions of an impersonal natural world and more as consciously self-aware beings involved in a network of interactions with others to whom they are accountable. The more they are seen as individuals, the more they are assumed to be individually morally responsible for their actions. Thus, the ideal-type here is the psychotherapeutic sciences, which locate the source of problems in the individual person(ality) and treat them as matters of individual conduct and right attitude (Laing 1967). Interventions principally focus on changing the behaviour or attitude of the individual, in effect calling them to moral account for what are seen as their personal failings. Accordingly, it is they themselves who are to blame. Blaming focuses on the moral failings of individuals, as failures to act or think in the right way. In wider contexts, however, it is equally in keeping with this mode of moral reasoning that blaming can be directed at some other individual rather than the self. Then, it is the other's failure to act or think properly rather than your own that is the problem and their fault that needs to be called to account and corrected. It is then easy to see how a 'culture of blaming' may arise within this form of moral reasoning in which salvation resides in correcting another person's moral failing.

Conspiratorial

On the other hand, at this end of the axis of moral accountability, where people are held to be collectively at fault, the focus of blaming is not directed at the individual but the collectivity. Thus, the source of moral failure is not sited in the individual actor or personality, but in the collective body to which the individual is ascribed membership and which is held responsible for shaping and impelling individual thought and action. The individual is caught up in the group's collective sway and the ways they act and think are not matters for which they themselves can be held responsible or called to account. They are subject to collective forces above and beyond their control and they may have little choice over their situation or the way of life they lead. Ideal-typically, this is expressed in those human sciences like sociology that stress the structuring power of the collective order. Often, the collective forces at work are assumed to have some central moment defined by what is taken to be the crucial positioning of one or other specific group. Thus, although the analytical logic is such as to treat all individuals as caught up in the collective body, some group or groups are so positioned as to have greater capacities to influence its shape and workings than others. So, even though their actions crucially delimit the capacities of other categories structured in relation to them, they and their actions are also defined by the general category to which they belong (class, gender, race etc.). Thus, blame cannot be directed at individuals as such, only at forms of structuring category. There is then a blaming of culture from which salvation is sought by disclosing the hidden mechanisms and structuring power of the dominant group so that it may be transformed.

Conspiring in the Light

Thus, the human sciences can be understood as developments out of the tensions informing disenchanted mundane reasoning over the nature of suffering and the direction of blaming. Since these tensions arise from basic dilemmas of moral accountability, we should expect them to play out not only between, but also within these sciences in the form of internal divisions. As argued in Chapter 1, such dilemmas provide the means of thinking (Billig 1996) and as such they tend to reappear and ramify at every level of application (Billig et al. 1988). In sociology, for example, they are manifest in the familiar structure-action 'duality' (Giddens 1984), which is not just ontological but also moral in that it raises questions about where we assign ultimate responsibility for action and its consequences – specifically, do we assign responsibility to individuals or to collectivities. One of the objections to an emphasis on action is that it is overly voluntaristic, implying that individuals are free to do whatever they choose without constraints imposed by a structuring sociality. So, in the case of ethnomethodology and conversation analysis for example, if social reality is constituted out of the *in situ* practical reasoning of members, then there is no need to consider the role of structuring constraints imposed by powerful groups; for example, 'rape' is only 'rape' if it is defined as such by the participants involved and whether they choose to do so or not has nothing to do with patriarchal power (see the debate involving Schegloff 1997, 1998, 1999a, 1999b, Wetherell 1998, and Billig 1999a, 1999b; see also Wooffitt 2005: 158-185). On the other hand, one of the objections to an emphasis on structure is that it is over-deterministic, implying that individuals have no freedom to choose how to act and hence no moral responsibility (Wrong 1961). So, in the case of Marxism for example, if individuals are 'subjects in ideology' (Althusser 1971), then how are they to take any role in bringing about revolutionary salvation from the evils of capitalism and why should they bother even if they could?

However, this is not just a dilemma for sociologists but also for ordinary people trying to decide who or what to blame for the suffering in their lives. One direction of resolution tends toward the blaming of culture and in this we can understand the parallel between critical sociology and conspiracy discourse. Both are articulations of the resources of accounting for suffering available within disenchanted mundane reasoning, resolving the dilemma in the direction of finding fault with a collectivity. If there is a difference, it is that conspiracy discourse is more thoroughgoing in its commitment to ascribing moral responsibility. Conspiracy discourse attributes blame at the highest level of abstraction within the 'people + group' locus by assigning responsibility, ultimately for everything, to the machinations of, typically, a single, small group of immense power and pervasiveness. Such a group is held to be morally culpable in so far as it is knowingly and deliberately behind a range of activities often of a formally illegal character (or at least questionably so – but then this group makes the laws so they can change them to suit), but in any case in breach of certain normative expectations whether of an economic, political, or generally social nature. In consequence, this group is responsible for suffering, both that of some specific individuals – especially the believers who are most actively conspired against – but ultimately for everyone else, the public in general.

It is this unwillingness to compromise in its moral commitment that makes conspiracy discourse seem 'excessive' and 'extreme'. To see it as 'irrational', however, is to see it from a different moral perspective informed by a different version of the discourse of blaming. But it is neither 'abnormal' nor 'pathological', simply an articulation of the means of moral accounting in a disenchanted world, in which the modes of mundane moral reasoning are shaped by a godless spirit that still nonetheless seeks a prospective pathway to salvation. God, supposedly, no longer offers a satisfactory path, while the path to disenchanted nature is a hard road to travel on. An alternative is offered by locating the source of suffering in people themselves. But if we do so merely as individuals, we face the potential obstacle that it is we ourselves who are to blame and we then who must take responsibility for our own salvation. The problem with the culture of blaming is that we might find only ourselves at the end of the road, caught in the glare of an unforgiving spotlight. Much better – surely, far more rational – is the option of blaming a collective other. Who then is to blame? No one and all; or better still, a small shadowy group, forever hidden just beyond view in the penumbra of the glare.

Chapter 9

Categorizing Jack – the 'Mad Doctor'

The last two chapters have looked at the phenomenon of 'mysteries' as a feature of contemporary disenchanted mundane reasoning, arguing that they show two main things: first, that the disenchanted orientation that seeks to make the unknown known actually produces unresolved argumentation; and second, that this is informed by a moral discourse directed at accounting for error in ways that attribute blame. Thus, far from eradicating the basis of moral judgement, the disenchanted mode of reasoning has developed a rich and varied moral discourse that continues to flourish in the form of the human sciences themselves. This moral discourse, however, is also present within the wider culture, where it informs the practical sociological reasoning of mundane commonsense, as is apparent in both Fortean and conspiracy discourse. This shows that science may be taken up and utilized by ordinary people in the form of characteristic styles of argumentation directed at legitimizing and delegitimizing knowledge-claims and in relation to this attributing blame.

In addition, however, this also suggests that science, in the specific sense of activities undertaken by scientists, is viewed in relation to an existing moral order into which these are positioned. As argued in Part II in the case of religion, the relationship between rationalization and religiosity is dynamic and mutually informing, involving discursive interplay as people actively attempt to figure out what science means in relation to religious beliefs and vice versa. However, there is no reason to assume this is confined solely to contexts explicitly said to be religious, but that a similar dynamic might be apparent also in the broader mundane moral order. Thus, contrary to the standard view of rationalization which insists the moral discourse of modernity is reduced to a one-dimensional monologue intoning the voice of uniform instrumentalism, PMS proposes that not only are the activities of scientists judged on moral grounds, but science itself is used as a resource of moral assessment (Cameron and Edge 1979). This does not mean just that science is interpreted through whatever existing moral framework(s) inform mundane reason, but that the resources provisioned by science are treated as inherently moral in that they are used as bases on which to type people and judge action within an order of social evaluation – one that includes scientists themselves. Thus, ordinary people are actively involved in attempting to work out both where the scientist fits in relation to the existing framework of moral assessment, but also reworking this framework by inventively developing the resources science provides to construct new bases of moral judgement.

A way into this is provided by membership categorization analysis (MCA), which was briefly discussed in Chapter 1, where I disputed Miller and McHoul's

(1998) view that this invalidates analysts' 'speculating' about meaning. I will now develop this to argue that speculation is an inherent feature of the inferential form of reasoning involved in membership categorization devices (MCDs), which conforms to the rhetorical figure of the enthymeme and is therefore probabilistic. As such, it is both moral and inventive. Membership categories constitute a moral ordering (Jayyusi 1984, 1991), though not in the sense of a pre-given, pre-defined framework that is simply imposed on people or situations (Hester and Eglin 1997b), but rather as an available resource that people use to work out possible moral assessments through the activity of categorizing. In rhetorical terms, this involves invention, the creation of arguments using available rhetorical resources – the means of persuasion – to think with, about, and through the meaning of situations in their specificity and immediacy. This reasoning has to be *done*, and done *in situ* as an ongoing practical accomplishment. However, in keeping with the incomplete logic of the enthymeme, the form this reasoning takes is narrative, involving the construction of accounts serving to function as explanations. Thus, people construct speculative stories (Eglin and Hester 1999), the sense of which is assessed by narrative criteria including coherence and credibility (Fisher 1994). Thus, for any given action or observable event to be attributed to a specific category of member, it must be understood as credible that that category might undertake such an action or be responsible for the event happening – unless it is the very *in*credibility of such a category taking such action that is the story. In either case, the action must (be made to) fit the perceived character-type or *persona* (Campbell 1975) and thus make for a coherent story.

To illustrate this, I look at one case that continues the focus on 'mysteries' and that, at least as a matter of speculative inference by some people, also involves certain types of scientist – that of Jack the Ripper. Before I say more about this, however, the above argument needs some unpacking.

Membership Categorization Devices and/as Enthymematic Reasoning

In Chapter 1, I referred to Sacks's example of the child's story, 'The baby cried. The mommy picked it up.' He argues that our sense of the story comes from our understanding of membership categories (MCs) and the activities associated with them, that 'crying' is a category-bound activity (CBA) of 'babies' and 'picking babies up' is a CBA of 'mommies'. Sacks's analysis is far more detailed than this, but there is a plentiful literature summarizing his work and no need to recount it here (Sacks 1972, 1995a, 1995b, Hester and Eglin 1997a, Silverman 1998, Lepper 2000, Schegloff 2007). However, there are some matters to discuss to bring out the link to enthymematic reasoning.

First, there is a methodological issue to address regarding the status of MCA and my own use of it, the central concern of which is its speculative nature. According to Schegloff (1995: xlii; see also 2007), Sacks abandoned MCA because it encouraged analysts to speculate beyond the observable detail of members' own sense-making, as is apparent even from the child's story. Sacks (1972: 32) defines a MCD as:

> that collection of membership categories, containing at least a category, that may be applied to some population, containing at least a Member, so as to provide, by the use of some rules of application, for the pairing of at least a population Member and a categorization device member. A device is then a collection plus rules of application.

This raises the question of how we decide what 'collection' is in use on a given occasion. In the child's story, Sacks (1995a: 238) argues the 'baby' and the 'mommy' come from the collection, 'family', formulating as a 'hearer's maxim' (Sacks 1995a: 248) that where two MCs are hearable as coming from the same 'team' (like 'family') then they should be heard this way. By this maxim and similar rules, members resolve ambiguities about categories like 'baby', which may belong to different collections – in addition to 'family' there is also what Sacks (1995a: 239) calls 'stage of life' and an additional collection, 'romance' has been, somewhat tentatively, suggested by Silverman (1998: 80).

However, it is far from clear that ambiguities about the meaning of categories can be quite so easily resolved. In the absence of some further confirmation, it cannot be definitively ruled out that 'the mommy' is not the mother of the baby, or not from the same collection, or that neither are from the collection 'family'. Arguably, our assumptions about what these categories mean relies as much on our prior knowledge that this is a *child's* story (Sacks 1974) as anything to do with the categories themselves, as this delimits the likely sense intended; we use our commonsense knowledge of the category 'child' to interpret the categories used by such a category. So *who* is using a category is relevant to our understanding of its sense, which raises the troublesome issue of context to which I return shortly. In addition, the notion of 'collection' is vague, as indicated by Silverman's tentativeness regarding the term 'romance', so analysts might then presume they can read a collection as relevant even if not explicitly apparent. For example, the use of an occupational category might be taken to imply the collection 'class', or a category like 'Jew' might be taken to imply 'race' (see below), but whilst some members might read these implications, for analysts to do so is not necessarily warranted. Consequently, MCA as Schegloff (1995: xlii) put it, risks 'promiscuity', which is why Sacks abandoned it in favour of studying the sequential organization of talk-in-interaction, that is conversation analysis (CA).

However, Watson (1997) argues that sequential organization is itself dependent on analysts' commonsensical understanding of MCs, for example in the use of such categories as 'caller' and 'called' in the analysis of telephone conversations. The issue this raises of pertinence here is how we are to approach the study of such social phenomena as newspaper reports, which make considerable use of MCs (Sacks 1995a: 205-222, Hester and Eglin 1997c, Eglin and Hester 1999). Given the practical impossibility of documenting on any large scale people's *in situ* sense-making work in their reading of such things – and even more so if our interest is at least partly historical – unless we are to abandon study of this kind, we are to a large degree left with little else to rely on except our own commonsense understanding. Thus, so far as the ensuing analysis is concerned, for reasons that will become clear, I see no alternative but to admit to 'promiscuity' (nothing 'romantic' implied!).

This, however, brings us back to context, or rather 'decontextualization' (Hester and Eglin 1997b). The worries about 'promiscuity' express a sense of epistemological validity that both STS and ethnomethodology in its more 'radical' moments (Pollner 1991) have done much to challenge. As argued in Chapter 1, the analyst is always at the end of the chain of interpretation regardless of what strategies we might employ to present this as warranted by the 'data'. In Pollner's (1987) terms as seen in Chapter 7, attempting to grasp the 'object' gives only the grasping: whatever instruments of detection we use, we only have what they detect, which is not the phenomena but a representation (so CA is dependent on its transcription techniques) that we still have to make sense of (so CA translates talk-in-interaction into a technical language that would likely be largely unrecognizable to the conversationalists [Billig 1999a]). It is notable in this respect that Sacks (for example, 1995a: 113-125) often represented his project through 'machine' metaphors, and his references to 'rules' governing MCDs and the 'systematics' of talk-in-interaction (Sacks, Schegloff and Jefferson 1974) suggest similar decontextualized conceptions. Some ethnomethodologists (Hester and Eglin 1997b), however, argue that MCs should be studied in a more contextually sensitive way that attends to the specific features attributed to categories on specific occasions of use. So, rather than attempting to extrapolate general 'rules' of usage, we should focus on the details of the content given to particular categories as and when they are used, and thus how categories are worked to constitute 'culture in action'. However, whilst this helps to bring out the active sense-making work people engage in, they nonetheless do so in part at least by drawing on available resources of understanding consisting of repertoires of representation and the suasive means to support them. Thus, any specific representation of a given category is only one of a range of possible ways it might be represented; hence it is potentially disputable, as is apparent from MCDs themselves because they are enthymematic.

For Aristotle (1946), the enthymeme or incomplete syllogism is the characteristic of rhetorical reasoning that differentiates it from logic: where logic presents a statement of a particular ('Socrates is a man') and a generalized justification ('All men are mortal') followed by a deduced conclusion ('Therefore, Socrates is mortal'), the enthymeme provides only 'a statement together with a justification' (Billig 1996: 132) or leaves out premises (Gross 1996: 12), obliging the audience to fill the gaps and make its own conclusions. More loosely, enthymematic reasoning is probabilistic, extending an open palm to the closed fist of the ostensibly unarguable syllogism. Thus, the audience's inferred conclusion may have been what was implied, but the rhetor might dispute it, just as the audience might dispute that the justification warrants the inference. In the delightful term Billig (1996) adopts from the sixteenth century rhetorician Ralph Lever, enthymematic reasoning both exemplifies and invites 'witcraft', the inventive disputation of inferences and premises.

MCDs are enthymemes bearing these rhetorical features, as a couple of examples from Sacks will show. Sacks argues MCs are 'inference rich' because of the activities 'bound' to them and accordingly, actions can be explained by reference to categories. However, as Silverman (1998: 75) points out, this may lead to accusations of 'prejudice', so one way to avoid this is 'to use a category and let others construct the explanation'. He cites Sacks's example of the government of the Soviet Union once publishing a list of names of 'profiteers' with no further comment – except that all the names were Jewish. Thus, the inference left for the audience of ordinary 'comrades' to draw was that the named individuals were profiteers because they were Jews. Stated enthymematically, the list is a series of particular statements of implicit form, 'X is a Jew' with an implied generalization inferentially 'bound' to this that 'Jews are profiteers' and the stated conclusion, 'X is a profiteer'. Neither the initial statement nor the justification were stated explicitly, but they did not have to be because inferential reasoning could be relied on to work back from the conclusion using what 'everyone knows' to be true about the category 'Jew'. But because it is unstated, the inference is also defeasible.

Another example from Sacks (1995a: 113) concerning a call to a social service agency shows how this works in a more clearly interactional situation (A is an agency worker, B the caller):

> A: Yeah, then what happened?
>
> B: Okay, in the meantime she [wife of B] says, "Don't ask the child nothing." Well, she stepped between me and the child, and I got up to walk out the door. When she stepped between me and the child, I went to move her out of the way. And then about that time her sister had called the police. I don't know how she ... what she...
>
> A: Didn't you smack her one?
>
> B: No.
>
> A: You're not telling me the story, Mr. B.
>
> B: Well, you see when you say smack you mean hit.
>
> A: Yeah, you shoved her. Is that it?
>
> B: Yeah, I shoved her.

Sacks draws attention to A's question, 'Didn't you smack her one?' and further insistence that B is not telling the full 'story', wondering how A, in the absence of any other information could know this with such certainty that B eventually admits to some physical violence. Sacks argues that A inferred an act of violence occurred because the police were called and a CBA of the police is dealing with violent incidents. Sacks (1995a: 116) spells out A's reasoning more fully:

> What we have is roughly something like this: A knows that the scene is "a family problem." So (a) is the family quarrel, (b) is the guy moving to the door ... (d) is the police coming. And (c) is the grounds for the police to have come. That is, apparently on some piece of information the police have come, and that piece of information is the thing that A has guessed at. A apparently knows, then, what good grounds are for the police to be called to a scene. And he's able to use those good grounds, first to make a guess, and then to assess the correctness of the answer to that guess.

As should be clear, A's reasoning rests on an enthymeme: given the initial premise of a family quarrel and the conclusion that the police were called, A fills in the missing step based on a general justification, the CBA that the police deal with violent incidents. Also apparent is the merely probabilistic character of enthymematic reasoning, as A's first 'guess' – that is, a probable supposition or speculation – is denied by B, who only accepts the validity of A's inference after 'smack', reformulated as 'hit', has been somewhat mollified to 'shove'. Thus, although A is correct in making the general inference that something happened for the police to be called, the precise grounds are only probable and therefore open to dispute.

It follows that like enthymemes, MCDs lack definitive closure – hence their 'promiscuity'. This was seen in Chapter 8 regarding the identification of Lee Harvey Oswald as President Kennedy's assassin and will be seen again shortly regarding Jack the Ripper. As Sacks put it (1995a: 42), people are continuously monitoring events 'to find out what is getting done by members of … categories'. In consequence, 'we get nicely special kinds of occurrences which provide a beautiful view of tensions arising as persons await the discovery of which of them is going to be found to have done this thing.' A pivotal point concerns the adequacy of 'fit' between a category and an individual's biography (Jayyusi 1984: 139-146) as this is always potentially disputable, either because some 'predicate' of the category does not fit the biography, or some characteristic of the individual does not fit the category.

The term 'predicates' refers to the full range of qualities associated with a category in addition to particular activities, which Hester (1998: 135) lists as: 'rights, entitlements, obligations, knowledge, attributes, and competencies.' Of notable significance to PMS is 'knowledge', as this implies that certain kinds of knowledge are 'owned' by specific categories, such as 'Azande witchcraft' or 'Baka medicine' (Sharrock 1974) – or indeed 'western science'. Sharrock argues such namings are not literally descriptive but refer to rights and entitlements (see also Sacks 1995a: 202-203) that carry moral significance. However, as demonstrated throughout this book, in the case of science, such rights are often disputed which follows from the incomplete, ambiguous and probabilistic nature of categories (see also Winiecki 2008). This is especially true of science as its own discourse enables it to be treated as both category-bound and category-free. It can be represented as both particularistic and universalistic; both 'science-in-particular' and 'science-in-general' (Michael 1992); both modern-western and transcendent of time and space; both 'privately' owned by the scientific priesthood and publicly shared by the synecdochic priestly voice; both alien to and aligned with commonsense.

Such ambiguity makes categorization always potentially arguable and resistant to closure. MCDs, as Silverman (1998: 89) states, are used by members to make '*presumptively* correct descriptions' – 'presumptive' because the description can be wrong and what people categorize using one collection might turn out to involve a different collection entirely. Thus, choice of category is significant and given there is always at least one alternative (Sacks 1972), categories do persuasive work by representing reality one way rather than another. Similarly, use of categories is 'recipient designed' (Silverman 1998: 89), chosen with a specific audience in mind to persuade them – and maybe also ourselves – to see things in one way rather than another.

All this dovetails neatly with the probabilistic nature of enthymematic reasoning, as also does a further feature Sacks (1995a: 42) mentions that the inferential nature of MCDs involves 'making new knowledge'. In linking predicates to categories and categories to predicates, people infer things about the ongoing events of the social world around them. But the lack of closure means there is always room for uncertainty and the potential that category ascription will be disputed. Thus, people have to be inventive in making category descriptions, both in the initial act of describing and to counter potential questioning of the categorization advanced. Categorization, then, is active work that may involve providing justification through elaborating both the description and the meaning of the category itself. Categories do not merely provide resources of narration as in the child's story, but are themselves developed and elaborated narratively as is perhaps most clearly apparent in the news media.

I will develop this point further below along with one other significant feature of categories, that they implicate a moral order as follows from certain of their predicates. Thus, as Jayyusi (1984) argued, categories are not just descriptive, but also evaluative and judgemental. Again, choice of category is critical in this: to describe someone as a 'Hell's Angel', for example, is to position them in relation to the prevailing categorical order and impute a set of behaviours already cast in a moral light, as 'good' or 'bad', 'praiseworthy' or 'blameworthy' and so on. None of this is definitive in its specifics as it depends on who is speaking to whom about what, but that there are judgements of value involved follows from the fact that an alternative category can always be used, such as one that might be neutral with respect to a collection like 'biker-gangs' or 'youth subcultures', but impute instead a different set of moral predicates (for example, 'bloke', 'mate', 'son', 'mechanic', 'lover' etc.) (see also Widdicombe and Wooffitt 1995, Widdicombe 1998).

This has been seen to some degree in earlier chapters regarding the category 'scientist', such as in the contrasting representations provided by positivism and Romanticism, which incorporate moral judgements about whether it is a good or bad thing to be a scientist, and about good and bad ways of doing science. MCA can help unpack this further through its focus on the active work of moral reasoning undertaken by people in their ordinary commonsense-making. In this, the category of 'scientist' is positioned within a pre-existing general moral order, but not in any simple, uniform or unproblematic way. Rather it is complex, mixed and puzzled activity as ordinary people try to figure out what 'scientist' means by defining, clarifying, elaborating – that is, applying – its predicates. This creative work is at once both descriptive and moral, involving judgements of 'badness' and 'madness' that are wrapped up together in narrative bundles that people unpack through enthymematic inferential reasoning. To demonstrate this, I now turn to the case of Jack the Ripper.

Why Jack?

Since nothing said above leads in any obvious way to Jack the Ripper, a word of explanation about this choice is in order. The reasons are primarily serendipitous (Ashmore, Myers and Potter 1995): I happened to be doing some work unrelated to either PMS or MCDs on the comic book (or 'graphic novel'), *From Hell* (Moore and Campbell 2000, see Locke 2009b), a fictional re-telling of the murders of several women in the Autumn of 1888 in the Whitechapel district of East London, commonly ascribed to killer or killers unknown referred to as 'Jack the Ripper'. In researching the topic, I noticed that the Ripper crimes are routinely represented as a 'mystery', leading to the line of thought developed in Chapter 7. This 'mystery' is not just a description of the 'fact' that there is no certainty about many aspects of the case including the Ripper's identity, but it is also their construal as such that gives the events their particular significance. Their status as 'mysterious' is used to warrant the interest shown by the 'Ripperologists' who have sought to identify the killer(s) and is central to the Ripper's signification in popular culture (Rumbelow 2004), but this status is itself an outcome of the events being oriented to in this specific fashion – as an unknown to be made known. So, while their status as a 'mystery' warrants the interest in them, this very interest confirms that status. Further, it is not new but was a theme in the early press reports of the murders (Walkowitz 1992: 196) that was echoed some thirty years later in the first book about them, entitled *The Mystery of Jack the Ripper*, and continues to resonate in contemporary 'Ripperature' (for example, Begg, Fido and Skinner 1991: back-cover, Knight 1994: frontispiece, and Rumbelow 2004: xv).

It might be argued that this simply shows that 'mystery' sells, but if so why does it? As I argued in Chapter 7, 'mystery' arises from the modern disenchanted form of mundane reasoning within which science plays a significant role. It is notable, then, that one projected solution to the 'mystery' has been the depiction of the Ripper as a 'mad doctor' (Frayling 1986). What is it about this type of character that enables it to provide such a solution? From within the orientation that seeks to make the unknown known, how is it that 'doctor' is able to provide for people a recognizably possible category to fill the 'known' slot? The obvious implication is that there is something about this type of character that is also 'mysterious' from the point of view of ordinary commonsense, which fits with a range of common images of the scientist whether as 'wizard', 'priest' or 'mad doctor'. But how exactly are such images used to construct accounts of events in the social world? What is the practical sociological reasoning involved?

These then are the questions directing my focus on Jack the Ripper. They provide a notable contrast to other literature on the topic, which is of two broad types: popular accounts that either offer projected solutions to the 'mystery' by identifying a candidate 'suspect' or review the 'evidence' for such (a small sample being Wilson and Odell 1987, Begg, Fido and Skinner 1991, Sugden 2002, Rumbelow 2004, Odell 2006); and historical and cultural studies situating the Ripper case in its historical context (Rubinstein 2000, Milburn 2008), contemporary media reportage (Walkowitz 1992, Curtis 2001) or broad cultural significance (Frayling 1986, Cameron and Fraser 1987, Tatar 1995, Smith 1997; see also Warwick and Willis 2007). A notable feature of some of the latter is that, although they avoid offering solutions in terms of identifying a specific candidate suspect, a general motivational explanation is advanced through the notion of *lustmord* (sexual murder), treating the Ripper as a case of male violence against women (see also Eglin and Hester 1999). Thus, the crimes are explained not biographically in terms of the blaming of an individual, but socially and categorically in terms of the blaming of culture (see Chapter 8).

So, these academic studies share with the popular literature a concern to proffer a motivational account and thus attribute blame. There is an obvious difficulty, however, in that the identity of the Ripper remains unknown and so any attempt to attribute blame, whether individually or categorically, is necessarily speculative. The Ripper, in Frayling's (1986: 214) apt term, is an 'absence'; to assign motive whether individual or collective is to fill this absence with doubtful presence. This applies also to the retrospective categorization of the Ripper as a 'serial killer' (Odell 2006) – a term coined over eighty years after the murders – and the use of techniques like 'psychological profiling'. Such forensic analysis is no more than the documentary method of interpretation (see Chapter 6) used to 'retcon' the Ripper into present forms of sense-making, treating history like comic book continuity by trying to fill it in to make one seamless, coherent narrative. A more specific problem with the attribution of *lustmord* is that sexual violence was among the prospective motivational accounts presented in the press of the time; as such, it is a resource of commonsense that needs to be studied (and it might also be pointed out that it has been speculated the Ripper was a woman [Odell 2006: 71-78]).

One significant feature is that in the unfolding speculative narratives about the Ripper, a prominent role was played by the category of 'doctor' as a type of scientist. Approaching the matter through MCA helps understand this while avoiding the above problems. As Sacks says, crimes provide moments of tension in which people wait to find out which category will be found responsible. In the interim, various prospective categories may be considered as possible culprits, such as 'sex murderer' or 'serial killer', and a 'profile' drawn up of their psychological 'type', that is a list of category predicates from which a specified category might be inferred. Thus, solving the crime in the first instance involves identifying potential suspect categories, then narrowing down the range until only one category remains, or perhaps one category set from which specific predicates can be triangulated to describe and match against individual biographies. The tension is only resolved, however, with the successful prosecution of a guilty party; in the case of the Ripper, no such prosecution ever occurred and so the tension remains – as Walkowitz (1992: 201) states, it 'offers no closure'. Instead, it has left a rich agar in which speculation about prospective categories has been able to grow uninhibited by the need to meet successful prosecution. As such, the Ripper case has much to tell us about the parameters of mundane reasoning involving the membership-categorical logic of criminality – and it is not insignificant that included within these parameters is the figure of the scientist.

This is apparent from contemporary newspaper reports of the murders, which I take from one particular secondary source. Here, I need to make some category entitlements clear. I am not a Ripper 'expert', but have relied completely on accounts provided by other scholars to whom I am greatly indebted. Most especially, I draw on Walkowitz (1992) as her account provides details of terms used in press reports of the time and for these I follow parts of her text closely. Nonetheless, my intention is to show that the data she documents can be reinterpreted to useful effect using MCA. It needs to be emphasized that my account is partial and selective, with particular focus on the categories of 'doctor' and 'scientist'. It should not be read as a full MCA of 'Jack the Ripper', but rather as an initial attempt to explore some features of the inferential reasoning involving these two categories. My use of MCA is also inspired by the above argument that it is essentially about rhetoric and, as befits the lighter touch of the open palm, I do not consider fruitful scholarship to be bound (as it were) by myths of strict methodological procedure (Billig 2004). The re-analysis presented here casts the data in an interesting alternative light that is productive for the general argument of this book and suggestive that further MCA of the categories 'doctor' and 'scientist' would provide valuable insight into PMS.

Walkowitz (1992: 225) herself views the press reports through the categories of 'class' and 'gender', stating that the 'Whitechapel horrors provoked multiple and contradictory responses, expressive of important cultural and social divisions within Victorian society'. She sees the press of the day as a forum for the articulation of the public discourse of the 'respectable' classes, specifically the male cohort, which she refers to as the 'Self', in contrast to an 'Other' consisting of women and certain residents of Whitechapel. Thus, as more murders occurred, she (1992: 206) argues that speculation about the identity of the Ripper crossed 'from an externalized version of the Other to a variation of the multiple, divided Self'. Given this, it is even more interesting that she (1992: 212) concludes that the 'mad doctor' has been 'the most enduring and publicly compelling member of a cast of privileged villains'. In other words, in her terms, it is 'Self' rather than 'Other' that remains the speculative guilty party.

But these terms are questionable and point to the issue raised above about the meaning of 'collection'. For Walkowitz, the relevant collections are 'gender' and 'class'; hence, the 'Self' for whom the press speaks is male and respectable, and because she includes doctors in this collection, she concludes this 'Self' was divided. However, it is not clear from her data that doctors are uniformly represented as part of this 'Self'; rather, they seem themselves to be viewed as an 'other' marked by a moral character that is uncertain and compromised. Critical to this is the ethic of conduct that according to the standard view of rationalization the scientist ideal-typifies: instrumental action. Thus, MCA suggests that, contrary to this view, instrumentalism does not erode the existing moral order, but is judged in its terms as a 'mad' and/or 'bad' 'other'. In addition, even where the resources of science *are* used as a basis of sense-making, categories are derived that are also given moral weighting and used with ironic witcraft, to tell possible 'atrocity stories' (Dingwall 1976) about doctors as men of science.

Jack-in-the-Press

As stated, 'Jack the Ripper' is an absence – a name we all know for someone nobody knows. It is a name of convenience that was taken from one of the numerous letters the police received during the course of their investigations and is thought by some Ripperologists to have been the invention of an unknown journalist (Rumbelow 2004). The name, then, might be taken to represent not the murderer so much as the murders and the surrounding activity they stimulated, including the press reportage which was extensive and filled with speculation from both writers and readers about the killer's identity. Part of what is not known is the number of murders that should be attributed to the same hand, but much of the literature focuses on the so-called 'canonical five' (or 'Macnaghten Five' [Odell 2006: xvii]) and Walkowitz's account deals mainly with these. They are, in order of date and time of discovery: Polly Nicholls, Annie Chapman, Elizabeth Stride, Catherine Eddowes, and Mary Kelly. They show what is taken to be a common *modus operandi*, involving the nature of the victim, the time of the murder, the weapon used and the form of injury. All five are said to have been prostitutes, a membership category that itself does particular kinds of work. They were each murdered during the early hours of a weekend morning (Stride and Eddowes on the same night) and in each case a knife was used with a blade sufficiently long to eviscerate and sharp enough almost to decapitate. In most cases, internal organs were removed from the body.

The widespread reportage of the murders incited expressions of great shock and outrage, with demands to find the perpetrator(s). As stated, Walkowitz argues that the press constituted a public sphere to which a wide body of citizens contributed, although largely confined to the male members of the 'respectable' classes. We need to remember, however, that letters are selected for publication by editors; thus, it is an open question as to how representative published views were even of such ‘respectable’ opinion. With this qualification in mind, press discourse can be viewed as a context where some of the culturally available possible MCs bound to acts of murder could be articulated, thus providing some indication of the boundaries of commonsense inference-making. The coverage was roughly split into two phases: one following the murder of Polly Nicholls, when attention initially focused on what might be called ‘the usual suspects’ local to Whitechapel; and a second following the murder of Annie Chapman, when the net was widened to include possible ‘respectable’ suspects. After Chapman’s murder, Walkowitz (1992: 205) describes the press reportage as more ‘sensational’ making references to ‘“bloodthirsty” monsters and fiends’. Thus, the categories used became more unusual and extreme, drawing on what she (1992: 196) calls the resources of the ‘fantastic’. This suggests that the boundaries of ‘normal deviance’, so to speak, had been breached and that the lack of closure enabled and encouraged speculation to stretch beyond established limits of the known to invent new grounds of the knowable. The category ‘doctor’ figured prominently in this speculative territory.

In the first phase, initial suspicion from both police and press was directed at possible ‘local’ suspects, that is category types (Jayyusi 1984) to which the activity of murder might be bound that were also directly linked to Whitechapel. Whitechapel itself was ‘a notorious, poor locale’, which for the ‘respectable ... public’ was ‘an immoral landscape of light and darkness, a nether region of illicit sex and crime’ (Walkowitz 1992: 193), often represented as a ‘jungle’ (Curtis 2001) populated by ‘savages’ (Rumbelow 2004: 5; see also Walkowitz 1992: 197). Thus, amongst the first suspects were ‘street gangs’ (Walkowitz 1992: 202) known to the police for extorting money from prostitutes and already suspected of two prior murders, those of Emma Smith and Martha Tabram (Rumbelow 2004: 29-33). Other suspects included further category types that might both be found in the environs of Whitechapel and have access to the sort of weapon the Ripper was assumed to have employed, such as butchers, shoemakers, and sailors off cattle boats.

One further category perceived by some to fit the frame even better was that of ‘Jew’, given specific characterization after the second murder as ‘Leather Apron’, following the discovery of such an item near Chapman’s body. Jews were amongst the more recent immigrants to Whitechapel and their involvement in a variety of trades using knives and other sharp tools – ‘Leather Apron’ was specifically described as a ‘slippermaker’ (Walkowitz 1992: 203) – combined with anti-Semitic sentiment to make ‘Jew’ a fitting suspect for many. Other speculations about Jews included ‘religious fanatics’ enacting a purported ‘blood ritual’ and ‘revolutionary socialists’ suspected by the police of spreading anarchy (Walkowitz 1992: 204).

In the second phase, a different set of category types was advanced. After Chapman’s murder, the sense that a different order of deviance was involved arose after details of the injuries to her body were presented at inquest. The police surgeon (quoted by Rumbelow 2004: 42) suggested that ‘the work was that of an expert – or one, at least, who had such knowledge of anatomical or pathological examinations as to be enabled to secure the pelvic organs with one sweep of the knife’. Press speculation then turned to categories of ‘expert’ who might possess such knowledge, notably types of doctor. Thus, whereas in the first phase there had been an insistence that 'no Englishman could have perpetrated such a horrible crime' (Walkowitz 1992: 203), it was now suggested that only an 'upper class' and 'cultivated intellect' (Walkowitz 1992: 208) could be responsible. One source of speculation came from *Doctor Jekyll and Mr. Hyde*, a stage version of which happened to be running in London at the time of the murders that presented Hyde as sexually motivated, a monstrous and violent 'manifestation of Jekyll's lust' (Walkowitz 1992: 206). The suggestion was then made that the Ripper might be 'an evolutionary throwback' or 'troglodyte' driven to violent murder by 'erotomania', a notion recently popularized from Krafft-Ebing's casebook of sex offenders, *Psychopathia Sexualis*. The Ripper was also cast as a 'crazed biologist' or 'mad physiologist' (Walkowitz 1992: 207), while one further notion was of 'syphilitic madness', with the speculation that the Ripper was a 'medical student' avenging himself on those from whom he had contracted the disease (Walkowitz 1992: 210). Significantly, however, sexual motivation was only one source of speculation and the papers 'debated whether the murderer ... was mad or vicious, a victim of disease or a practitioner of "mere debauchery", a "homicidal maniac" bent on violence or an "erotomaniac" bent on sexual satisfaction' (Walkowitz 1992: 208). Further proposals were 'human vivisectionist' and 'bodysnatcher', while another type of medical motivation, suggested by the coroner at Chapman's inquest and described by him as more 'rational', was that the Ripper was harvesting a uterus to sell as there was 'a market for the missing organ' (quoted Walkowitz 1992: 209). There were also other types of 'privileged villain' proposed, including one other type of scientist, 'the Scientific Sociologist' (Walkowitz 1992: 212).

Overall, Walkowitz (1992: 212) argues that 'these speculations oscillated between two expert discourses: the language of the law, emphasizing free will, responsibility, and reason; and the language of medicine, which focused on nature, determination, and irresponsibility.' It will be recalled from Chapter 8, that Dean (1998) noted a similar contrast in the discourses informing the debate about UFOs and alien abduction, a tension between the 'democratic-legal' view of the abductee as rational citizen and the 'scientific' view of them as irrational subject. In the former view, people have free will, the capacity to choose how to act and so bear moral responsibility for their actions; in the latter, their (our) behaviour is determined by forces beyond our personal control and we cannot be held morally responsible. What the Ripper case then provides is a fascinating example of this dilemma being used as a means of thinking about the category of the scientist in terms of the specific predicates bound to this. Accordingly, the scientist is subject to the available means of moral reasoning rather than science displacing this with disenchanted instrumentalism. This can be brought out more fully by closer consideration of the category work involved.

Jack the MCD

As seen, in the first phase, suspicion was directed at 'usual suspects' or 'ordinary' deviants. It is this that indicates that the murder of Polly Nicholls was not considered anything especially remarkable, only the latest in a string of murders of 'prostitutes' in a district that was already represented amongst the 'respectable' public as both deadly and debauched. Thus, 'prostitute' – and perhaps more so 'Whitechapel prostitute' – was an MCD that carried a specific moral predicate, 'women of evil life' (quoted Walkowitz 1992: 218) enabling some to draw the inference that 'she brought it on herself', or even 'she got what she deserved'. In the first instance, then, neither the police nor the press treated this murder as anything particularly special as murders go and their inferential reasoning directed them towards MCs, the commonly known activities of which might put them in the frame. However, even amongst these were those typically known as 'bad', such as street gangs, and others considered potentially 'mad', notably the 'Jew'.

As Sacks (1995a: 180) commented about certain categories like 'Jew', for some, this alone would be enough to make them suspect. Notable in the Ripper case was the contrast with the category 'Englishman': whereas the latter was explicitly considered incapable of 'perpetrating such a horrible crime', no such limitation applied to the former – a measure then of the category's perceived difference in terms of its normative boundedness. The collection these two categories come from then might be 'ethnicity' (Day 1998), though perhaps more appropriate to the time is 'race'. (It might be suggested that 'Englishman' is a gendered category, but although this seems obvious in the early twenty-first century academy, whether it would have been considered so by the late nineteenth century 'respectable' public is less clear.) However, although the category 'Jew' alone might have been considered sufficiently probable to direct suspicion, additional category predicates were looked for to bolster the inference by strengthening the chain linking category to biography. Here, a different collection provided additional resources, 'occupation'. This is how we can understand the suspicion directed at 'Leather Apron', as 'slipper-making' was an occupational activity bound to 'Jew'. This could then be used to account for the types of instruments and skills shown by the Ripper, as well as the discovery of the leather apron. However, this still might not be considered sufficient justification, as whether 'Jew' or not, an activity not commonly associated with 'slipper-maker' is 'murder'. Hence, further category predicates were also sought. In the case of Jews, two other collections available were 'politics' and 'religion'; thus, while the police speculated about 'socialists and revolutionaries' attempting to ferment popular unrest, the press speculated about a 'religious fanatic' atoning for sexual 'pollution' (Walkowitz 1992: 203).

In this, then, we can see something of the way inferential reasoning is used to 'make new knowledge', filling in categories through connecting their predicates to construct possible narratives to account for events and so figure out ways to link categories to biographies. The kinds of connections looked for are specific to the categories in terms of their commonly recognized predicates, which lends them narrative coherence so that a *presumptively* credible story can be told. This does not make it true, but for some at least it might seem probable. Nonetheless, whilst speculation built around the category 'Jew', it remained within the boundaries of presumptively (however unjustly) 'known' deviants. Where the Ripper case becomes more fascinating is during the second phase of speculation, when the boundaries of known deviants began to be stretched to try to accommodate the 'monstrousness' of the crimes. Here, a different set of MCDs was deployed, the atypicality of which shows that the events were now being actively (re)constructed as something extraordinary. Speculation moved beyond the 'usual suspects' to encompass a much wider range of candidates, seeking out categories of member to which such extraordinary activity might be bound, to find 'monsters' – and amongst these was the category 'doctor' conceived as a man of science.

It is not insignificant that the speculative potential here was first opened up by a member of the medical profession, the police surgeon at Chapman's inquest with his suggestion that her wounds showed an 'expert' hand at work. Such a judgement coming from a doctor presented the possibility of inferential reasoning along the lines of 'takes one to know one'. That a skilled practitioner was declaring the activity to be recognizably skilled opened the possibility that the murderer was from the same category. The work of creating new knowledge then focused on generating speculative narratives in which a doctor could play the role of villain. As it happened, one was already to hand in the form of 'Dr. Jekyll' and all inferential reasoning needed was to pull aside the stage curtains, allowing Jekyll's monstrous alter ego to step into the streets of Whitechapel. This highlights another way in which the boundary between 'fact' and 'fiction' can become blurred (see Chapter 6). The Ripper's crimes tore the fabric of the normative order to such an extent that the boundaries of the 'real' became uncertain and the impossible began to seem, not just possible, but perhaps even probable. For some at least, the fictional imaginings of the possible activities of scientists began to seem quite reasonable inferences.

However, science also provided additional resources with which to construct narratives of 'madness' and 'badness' of a less obviously 'fictional' form. The varieties of 'mad doctor' advanced as putative identities for the Ripper included: 'evolutionary throwback/troglodyte'; 'homicidal maniac'; 'erotomaniac'; 'vengeful syphilitic'; 'body-snatcher'; 'vivisectionist'; and 'scientific sociologist'. How do these categories work as projected resolutions to the mystery? As with the category 'Jew', there is a need to link category to biography and, as with the category 'slipper-maker', the suspicion that a 'doctor' might have been involved raises a problem. Typically, the occupational category 'doctor' is no more ordinarily bound to the activity 'murder' than is 'slipper-maker', so just as with 'Leather Apron' inferential speculation was led to consider further predicates possibly associated with 'doctor'. In the case of Jews, predicates drew from the collections 'religion' and 'politics', providing varieties of 'mad' to add to the 'bad' of ordinary criminals. The equivalent question with respect to doctors then is: what kinds of 'mad' and 'bad' might be associated with them? The prospective answers draw from recognizably known things about scientists in two different ways: one uses the stories scientists themselves were beginning to construct about the nature of humanity in general, to apply these with ironic wit to the category 'scientist'; the other draws on predicates specifically bound to the category 'scientist' with one common characteristic – instrumentalism.

In the first case, then, we have 'evolutionary throwback'. This explicitly draws on the scientific view that human beings have evolved from 'primitive' species of hominid and then enables the speculative notion that prehistoric 'troglodytes' might somehow reappear as a 'throwback' of reproduction or result from some bizarre scientific experiment. Given that the populace of Whitechapel were themselves likened to 'savages' by 'respectable' commentators, it became possible to imagine such a 'primitive' appearing even amongst 'cultivated Englishmen' themselves. Thus, I would suggest, this category draws from the collection 'race' made available by the discourse of science and thus has some comparability to the category 'Jew' as a basis of inferential reasoning. Both categories provide means of defining 'otherness' on the basis of species-membership and so implicate a moral order in which species are ranked in terms of their presumptive level of moral development.

Meanwhile, the categories 'homicidal maniac', 'erotomaniac' and 'vengeful syphilitic' use a different means of defining 'otherness', but also draw on an emerging scientific discourse: psychopathology. This defined 'madness' in medical terms and pursued systematic classification of its forms (Foucault 2001). Thus, just as the notion of 'erotomania' was taken from Krafft-Ebbing, so was the idea of a specific connection between venereal disease and madness taken from current medical science (Walkowitz 1992: 210). These categories, then, draw from a different collection, 'mental illness', but one again made available by the discourse of science. This implicates a moral order directed at an individual level rather than a whole species, providing a different basis on which to judge action with different possible consequences and outcomes, as well as alternative narrative scenarios.

However, although they implicate different kinds of moral order, both the collections 'race' and 'mental illness' partake of science's universalistic rhetoric and as such are ways of categorizing everybody, not just scientists. Thus, applying them to doctors shows ordinary people's inventive witcraft, turning the tables on scientists by using their own rhetoric to cast moral judgement presumptively on a suspected member of the scientific community itself. The remaining categories, however – 'body-snatcher', 'vivisectionist' and 'sociologist' – show a further difference, as they refer to activities specific of or directly linked to scientists. 'Body-snatchers' procured human corpses to sell to medical researchers and the suggestion was made that the Ripper was doing likewise with female organs; similarly, 'vivisectionists' practised their research on live animals and the horror of the Ripper was the possibility that he might be doing the same to women. In either case, activities bound to medical researchers were being drawn on, activities of an essentially instrumental nature in their treatment of human beings. Whether the Ripper was reaping organs for cash or knowledge for medicine, he was treating women as a means to an end. The 'scientific sociologist', too, shows a similar morally compromised instrumentalism. The category, proposed by Bernard Shaw 'facetiously' according to Walkowitz (1992: 212), was described as 'a social reformer trying to expose the conditions of the East End'. Facetiously or not, it is significant that it *could* be proposed as a CBA of the scientist, treating ordinary people's lives as a means to an end even if with 'good' moral intent. Thus, the collection for these categories is 'scientist' and forms of instrumental action are viewed as a CBA of this category. But this is not just a description, it is also a moral judgement in which the 'scientist' appears as an 'other', a category outside the existing moral order and as such of uncertain and potentially dangerous status. The scientist's instrumental 'ethic', therefore, is both category bound and bound to no good.

Beyond Instrumentalism – Towards a Conclusion

It must be stressed again that this analysis is provisional and intended to be no more than suggestive of the possibilities of MCA for studying the category 'scientist' in PMS (see also Winiecki 2008). In this respect, there are three main points to stress.

First, it provides some support for the more general argument made elsewhere (Locke 2002) that PMS is rhetorically inventive, particularly involving enthymematic reasoning. The above analysis shows that people can and sometimes do draw on science as a resource, using it to construct prospective descriptions of action that then have a 'scientific' character to them, albeit of a speculative and hence arguable form. People do use science to construct understandings of actions and events in the social world; however, these understandings may or may not conform to those currently advocated by members of the scientific community (itself speaking a multiplicitous discourse). The reason for this is not necessarily a matter of 'ignorance' or other form of 'deficit', but because people use science as one source of materials to fill in the gaps to account for observed actions and thereby prospectively seek to resolve the tension in finding out what category did the thing. Accordingly, science is treated as a resource of moral reasoning, a means with which to construct categories of potential actor, who are always already positioned morally by virtue of the action to which they are linked. Thus, the relationship between science and the existing moral order(s) is dynamic and dialogical: science provides new categories to build into the existing moral order which becomes reconfigured accordingly; but the order remains a moral one and the scientific categories are positioned within its system of relative valuations. This dynamic is fuelled by the inventive use of science to tell stories about both the possible actions of known categories and the possible categories behind known actions. This is done with witcraft, using ordinary skills of rhetorical reasoning to make a persuasive case for perceiving the matter one way rather than another. A notable feature of this witcraft in the case of Jack the Ripper was to use the universalistic rhetoric of science against scientists, cleverly arguing that a prospectively culpable man of science was himself suffering the effects of a scientifically conceived degeneracy, whether of a species-wide or individual nature. Whatever such stories may have lacked in scientific credibility they made up for in their ironic narrative coherence.

Second, the 'scientist' as a category may itself be represented as a moral 'other' by virtue of a form of activity bound to it: instrumental action. Such action is a focus of moral opprobrium through a variety of negatively evaluated category types linked to it, all of which are directly associated with the doings of scientists. This is not to argue that this is always the case and it may be that in some contexts acting instrumentally is given positive moral assessment as cause for praise rather than blame. Nonetheless, the significant point is that it can be a basis of negative judgement and that it is morally judged at all. Instrumental action is not, therefore, outside of moral assessment – in being conceived as amoral, it is thereby judged morally. Thus, if this is treated as a predicate of the category 'scientist', then this category is assessed in relation to existing moral standards. Accordingly, science does not necessarily or straightforwardly displace or dissolve these moral standards. Rather, the relationship is dynamic and scientists may be held accountable by and to a moral order that, while lacking formal codification, is that much more powerful by virtue of its informality and mundane constitution. In relation to this, the scientist stands as an uncertain, ambivalent and somewhat mysterious figure, both somehow known and not known. The scientist, then, is 'other' regardless of 'class' or 'gender' or 'respectability', because these are not the relevant collections so much as 'occupation'. In the Ripper case, when the going got 'monstrous', the monstrous was got going – specifically the occupational category 'doctor' conceived as a man of science. For mundane categorical reasoning, 'doctor' as a category is conceivable as 'monstrous' in ways that other aligned categories are not. It is important to recognize that this is about *categories*, not biographies; the categorical order has its own workings regardless of the individuals that fill category positions. Exceptions can always be made without calling the categorical order into question (Jayyusi 1984; see also Billig 1996 148-185); rather, the categorical order is used to judge individuals as 'deviant', as 'bad' or 'mad'. Occupations are part of a categorical order to which all adult members belong (even if only by virtue of being 'unemployed' or 'retired') and within which every occupation is a potential 'other' to *any* other (Parker 1995) – so 'journalists', for example, might readily consider 'doctors' quite 'other' to themselves. Each occupation has a different set of predicates attached, enabling different inferences to be drawn and thus potentially different stories to be told about each. Thus, the categories at work need to be studied in their specifics to understand what predicates are employed and in what ways, and how then the 'doctor' and/or 'scientist' is morally positioned and re-positioned on each occasion – 'each next first time' (Garfinkel 2002).

Given this, it might then be argued that there is little to be learned about contemporary views of either the 'doctor' or the 'scientist' from the Ripper case and that, so far as disenchantment is concerned, it pre-dates the time when the process of moral displacement really got underway. To the contrary, however, the Ripper case is significant, because the view of the culprit as a 'mad doctor' continued to be played out in popular culture over the course of the twentieth century (Frayling 1986), as *From Hell* illustrates (Locke 2009b). Moreover, there are grounds for arguing that 'mad doctors' have become both more common and more extreme figures in popular depictions of the scientist (Skal 1998). Here, however, the point about specificity does count, as we should be wary of assuming that every such figure is more or less the same (contrast Basalla 1976). So, while the Ripper case provides some support for the points above, these need to be treated as speculative generalizations providing possible bases for further enquiry into the empirical detail of other cases. They encourage an approach to such enquiry using MCA that avoids assuming we already know what predicates the category 'scientist' (and/or 'doctor') has bound to it, or that any one use is the same as any other, but instead focuses on specific usages in specific contexts to begin to build up a more informed and detailed picture of mundane moral reasoning about science. Seeing the 'scientist' in instrumentalist terms is only one dimension of this mundane moral reasoning; as such, it is itself a rhetorical construct providing only one enthymematic narration of the public meaning of science. This contributes to PMS, but public and popular inventive wit reaches far beyond.

Not Quite a Conclusion

In this book, I have sought to make a case for an approach to the study of public (and popular) meanings of science through rhetorical sociology. This has been constructed through a re-crafted framework of rationalization that, although grounded in Weber's analytical model, attempts to recover from this a wider range of possible forms of PMS than just a monolithical condition of disenchantment. From this, I have distinguished five aspects of PMS:

1. *The standard view of rationalization.* Here science is both disenchanted and disenchants. Disenchantment is understood as involving both theoretical mastery (abstraction, universalization and depersonalization) and calculation of means (an instrumental orientation) – or roughly, rationalism and empiricism respectively. Thus, scientists present a self-image of disinterested, rational knowledge accumulation and science encourages a discursive mode in which unknowns are oriented to as in principle and potentially knowable, especially through the experimental way of life. This is a major component of the formal public rhetoric of science and is prominent in scientists' self-presentations directed internally toward the scientific community.
2. *Enchanted science.* Here, the unfolding discourse of the charismatic is articulated in forms of theoretical mastery imbued with a spirit of complete cosmological comprehension that offer synecdochic visions of universal wonderment and practical empowerment (especially in the adulation of technology) – the grand metanarratives of science. These are also part of the formal public rhetoric of science, especially directed towards external social groups for purposes of maintaining legitimacy and professional reproduction through recruiting new adherents (the 'priestly voice'), but they may be taken up in unexpected ways and used in the construction of alternative cosmologies seeking to extend the charismatic reach beyond science.
3. *Critical resources of argumentation drawn from science.* Broadly, there are four of these:
   a. The disjuncture between theoretical mastery and calculation of means expressed in enchanted and disenchanted versions of science enables alternative visions critical of instrumentalist reductionism to be invented, articulated and legitimized, in terms consonant with scientists' own discourse. This type of critique informs the aesthetic sphere, especially Romanticism as is apparent in science fiction and fantasy. It is also found in the discourse of reflexive spirituality and some forms of fringe science including aspects of Forteanism.

b. 'Formal irrationality', or cognitive and methodological errors. These are the critical counterparts to rationalism and empiricism, providing scientists with bases of error accounting that may be employed to undermine alternative versions of reality. However, proponents of those versions may employ them to equivalent effect against 'orthodox' or conventional science, producing a range of unresolved arguments in a growing and thickening scientific fringe.
c. 'Substantive rationality', or social interests. A form of scientists' error accounting that implicates the presence of general ethics of conduct drawn from other value spheres interfering with the 'purity' of the scientific process, such as economic, political or religious interests. Again, this is employed both for and against 'orthodoxy', helping to thicken the rhetorical range of the fringe.
d. 'Substantive irrationality', or personal factors. A form of scientists' error accounting that implicates the presence of particular characteristics of the individual interfering with the purity of the scientific process, again employed with equivalent effect by the fringe.

4. *Mundane mysteries (disenchanted mundane reason).* The disenchanted orientation directed towards making the unknown known has the paradoxical consequence of producing irresolution, especially focused around anomalies and mysteries. In conjunction with scientists' means of error accounting and the spirit of Romanticism, this produces a sharpened politics of experience built around the contrast between 'skeptical debunkers' and 'believers' found in Fortean discourse. But it also produces a 'third way' in the form of 'lay' (that is, 'non-scientist') enquiry into 'strange phenomena', with 'sceptical' uncertainty expressed in the Fortean 'Great Object' as a mirthful Trickster.
5. *Membership categorization and the moral order.* Here, the category 'scientist' is viewed for its presumptive predicates both as a resource of new moral categorization and for its 'bound' activities in relation to an established moral order. The category 'scientist' may then be judged either through the terms of scientists' own discourse being wittily turned against them or by their presumed instrumental activity used to find them morally wanting.

Broadly, these different aspects are developments out of a central dilemma concerning the meaning of science within modern culture: does it disenchant or enchant? This is expressed clearly in the contrast between the standard view of rationalization and the cosmological visions provided by the 'grand metanarratives', but it informs the other aspects as well. It provides resources of critical argumentation by presenting a de-moralized ethic of conduct for the scientist, ideal-typified as a pure instrumental actor, critically contrasted with deviations that constitute varieties of 'formal' and 'substantive' error employed by 'debunkers' as means of boundary-maintenance. Paradoxically, however, these contribute to the means of mystery maintenance, as they provide resources with which to defend alternative versions of reality against scientific 'orthodoxy', some of which draw on enchanted views of science. Meanwhile, the ideal-type is also a focus of critical characterization of the scientist as a membership category bound by forms of instrumental activity that are judged morally wanting. Thus, the dilemma of dis/enchantment is cross-cut by a second dilemma concerning the moral evaluation of scientific action as 'good' or 'bad' in either case.

Conclusion 1 – W(h)ither Charisma?

This analysis leads to two rather different conclusions. In one, the logic of the account is followed through to consider the further prospects of intellectualist rationalization. In Weber's model, the unfolding discourse of charisma develops from critique of the established social order in response to the existence of suffering, which is led by charismatic virtuosos finding flaws in the legitimizing belief, enabling them to construct alternatives with more encompassing social appeal. This has resulted in progressive abstraction and universalization at the level of theoretical mastery, which in the west combined with a world-transformative ethic to produce a growing instrumental orientation – the condition of disenchantment associated with modern science and technology.

However, it is apparent from the above that disenchantment does not disable the potential for charismatic critique of science, because the dilemma with enchantment produces resources to advance and defend alternative cosmologies. These arise from the disjuncture between theoretical mastery and calculation of means, and the forms of error accounting that scientists themselves employ. In combination, these produce two contrasting outcomes: one is a higher level charismatic reaching beyond instrumentalism towards a more encompassing abstract universalism that synthesizes science with moral orders both religious and mundane; the other is a state of irresolution that conflicts with the idealizations of disenchanted mundane reason to provide a determinate, coherent and non-contradictory Great Object. In short, they define a further dilemma between singular definitiveness and uncertain multiplicity – a 'closed' or an 'open' society.

This is rather different from a 'postmodern condition' of 'paralogy' in which the construction of new 'language-games' is an end in itself (Lyotard 1984), as that misses the dilemmatical nature of the situation in which discursive multiplicity is an outcome of the reaching for singularity. The grand metanarratives have not collapsed so much as diversified and hybridized, as the charismatic continues to seek a higher level universalism. Thus, the dilemma is precisely that there is available a multiplicity of contenders each of which claims to provide the singular definitive truth – and it is sharpened by each of them mobilizing the rhetoric of science in support. One possible outcome of this is scepticism towards science, but this should not be equated with a singular condition of 'risk' or 'anxiety' as this misses the nature of the dilemma: it is the belief in the in principle knowability of the world that sustains the condition of uncertainty. This may indeed encourage critical enquiry and reflection, but this is no more than saying that arguing encourages thinking (Billig 1996) – and for the critical thinker, 'risk' can become 'opportunity'. Thus, if there has been a growth and widening of critical thinking about science, then this provides fuel for charismatic virtuosos to construct and articulate their alternative visions. Again, therefore, the charismatic continues its reaching into abstraction even as the available versions multiply.

The notable feature of the versions of charismatic reaching discussed in this book is that they share, albeit in somewhat different ways, a focus around the constitution of the symbolic as argued by Whitehead (1974). So, this is seen in attempts to construct discursive syncretisms between religion and science, whether in aid of advancing fundamentalism or developing new forms of spirituality, as in either case symbolism is a central tool in the effort to forge metaphorical connections and articulate synecdochic reductions. However, in the discourse of reflexive spirituality, the interest in symbolization becomes more explicit through the attempt to detach symbols from their 'literal' groundings in established ethics of conduct – or we might say, to free signifiers from signifieds – which culminates in a view of reality as constituted through the magic of language. Something similar is also seen in the Fortean conception of 'strangeness', as a condition in which the understanding presented of the world is inextricably tied up with the means of understanding employed, so that all explanations including their own need to be treated as part of the cosmic joke. And it is seen again, if rather differently, in conspiracy discourse which treats every feature of the world as a sign pointing toward massive singularity, one that includes the forms of understanding employed by conspiracists and their academic critics alike.

Taken overall, these present four ideal-typified charismatic solutions available in contemporary culture as resolutions to the dilemma over the meaning of science as dis/enchanting cross-cut by the dilemma of singularity/multiplicity. Thus, on the side of 'singularity', where creationism/ID offers an enchanted view of self as part of the singular cosmological order mapped out in God's great Design, conspiracy discourse offers a version of disenchanted mundane reasoning's response to suffering that points the finger of blame towards a single, dominating shadowy group. Meanwhile, on the side of 'multiplicity', where reflexive spirituality offers an enchanted view of self as part of a cosmological order in which all visions of reality are accommodated within 'Ideaspace' (or as aspects of thetans' past lives), Fortean discourse articulates disenchanted mundane reason's 'skepticism' towards fantastical visions even as it cocks a 'sceptical' eye towards the 'skeptics'. Which, if any, of these solutions may prevail as the basis of an alternative ethic of conduct to disenchanted instrumentalism is an open question, but they provide some basis to begin to develop a fuller sociological understanding of PMS and the dilemmatical resources of thinking and arguing from which such an alternative may arise.

Conclusion 2 – Towards a Sociological Anarchism

But of course this is all too neat. Such a conclusion itself bespeaks only one side of the dilemma in telling a tale of singularity, one that runs throughout this whole text: the sociological grand metanarrative of rationalization. But if this text is to understand itself as a social object, an articulation of the sociological discursive and an account that like all accounts reflexively constitutes the context it presents as its situation – in short, as a rhetorical construct – then it must acknowledge the ironic character of its own thesis and that 'rationalization' is an outcome of the application of the documentary method of interpretation (or not, for this too is a grand metanarrative). The paradox of consequences of a thesis that rationalization produces an orientation towards the nature of symbolization is that it is obliged to recognize its own representation of reality as a symbolic construct.

So here is a second conclusion, one that ends appropriately enough where it all began, with words and the slippages of meanings. The attempt to stop words slipping away and to close down meaning has about it something occultic. In traditional magical belief, to know something's true name is to have power over it, to bind it to your will and make it do your bidding. Embedded in this is a profound social truth concerning the struggle for power: to know the 'true' name of things means to have won the battle over meaning, to have the capacity to assert definitively the way the world is and impose this way on others. This becomes an occult act – 'hidden; ... beyond the range of understanding or of ordinary knowledge' (OED) – because a central strategy in the struggle is to disguise the social origin of the definition, to make it appear natural, inherent and given, and so invest it with transcendence as an unchanging and unchangeable truth. But despite whatever efforts are made to stop them, words still slip from control as they transmogrify in and through their usage. Whatever magic tricks we might conjure from our rhetorical hats to try to pin them down, they just keep flowing away. The ship may be in the bottle, but it drifts away with the tide nonetheless.

For the sociologist, this calls for a different kind of magical thinking I call *sociological anarchism*. This also has parallels in ancient beliefs, but ones that lead us away from eternal verities towards embracing our own historical situatedness. In Norse mythology, the high god, Odin possessed two ravens, Hugin and Munin, usually translated as 'thought' and 'memory'. Every morning they flew from his shoulders to scour the world and on their return informed him of all that was and all that had been. Odin did not think he already knew what was going on in the world or what it was like; he had to have some messengers inform him. This is the attitude of the sociological anarchist: not to presume we already know what the world looks like, but to set about finding out with our two ravens, Recordin and Remindin. We record the struggles over meaning currently going on, documenting the signifying practices employed by the social groups involved, tracing the constructions of the social they configure in their categorizations, representations and rhetorics, even as these slip and flow away. And we remind about the struggles over meaning that have gone before through recovery of the social configurations built into existing meanings that might now appear as unchangeable verities, but which even as we recover them may already be slipping and flowing anew (Latour 1987).

A further task for the sociological anarchist follows, one that resonates with another ancient practice: *memento mori*. This is the phrase, so it is said, that slaves whispered to the heroes of the Roman Empire when fêted by adoring crowds to remind them they are mortal. It is a reminder that, however great we might think we are, whatever power we might currently command and however far above the ordinary run of humanity we might believe ourselves to be, our time too will pass. Mortal greatness and power will fall into the footnotes of history and eventually crumble to dust. So, the sociological anarchist mutters an equivalent whisper to those who emerge victorious in the struggle over meaning, *memento interpretatio*: 'Remember you are socially constructed!'

So equipped, the sociological anarchist lurks in the shadows of the social, ever prepared to toss metaphorical bombs at the signifying practices of the perpetrators of permanency to explode their myth-making. The sociological anarchist hugs the margins and avoids the centres of action, aware that any centre is only ever temporary and will soon seem just another margin made up to look a full page. But wary of forgetting their own socially constructed nature and wanting to avoid the self-delusions of fixity, the sociological anarchist always keeps back one bomb for themselves.

So it turns out that sociological anarchism is a form of Marxism – Groucho Marxism. Groucho famously remarked that he refused to join any club that would have him as a member. Taking this for his cue, the sociological anarchist will only join the club rejected by everyone else – including the members themselves. And so, in the final paradox of consequences, the sociological anarchist laughingly echoes the ancient Cretan and in all seriousness, stretches out her open palm declaring: 'All sociologists are liars!'

Bibliography

Aaronovitch, D. 2009. *Voodoo Histories: The Role of the Conspiracy Theory in Shaping Modern History*. London: Jonathan Cape.

Abercrombie, N., Baker, J., Brett, S. and Foster, J. 1970. Superstition and religion: the God of the gaps, in *A Sociological Yearbook of Religion in Britain 3*, edited by D. Martin and M. Hill. London: SCM, 93-129.

Abercrombie, N., Hill, S. and Turner, B.S. 1980. *The Dominant Ideology Thesis*. London: Allen and Unwin.

Adorno, T. 1991. *The Culture Industry: Selected Essays on Mass Culture*. London: Routledge.

Aldiss, B.W. 1986. *Trillion Year Spree: The History of Science Fiction*. London: Victor Gollancz.

Althusser, L. 1971. *Lenin and Philosophy and Other Essays*, translated by B. Brewster. London: New Left Books.

Amsler, S. 2007. Knowledge, freedom and post-Soviet imperialism: the case of social science in Kyrgyzstan, in *Theorising Social Change in Post-Soviet Countries: Critical Approaches*, edited by B. Sanghera, S. Amsler and T. Yarkova. Bern: Peter Lang, 171-98.

Anon. 2007. FT's opinion. *Fortean Times*, 227, September, 73.

Aristotle. 1946. *The Works of Aristotle Translated into English: Volume XI. Rhetorica*, translated by W.R. Roberts. London: Oxford University Press.

Aronowitz, S. 1988. *Science as Power: Discourse and Ideology in Modern Society*. London: Macmillan.

Ashmore, M. 1989. *The Reflexive Thesis: Wrighting Sociology of Scientific Knowledge*. Chicago: University of Chicago Press.

Ashmore, M., Myers, G. and Potter, J. 1995. Discourse, rhetoric, reflexivity: seven days in the library, in *Handbook of Science and Technology Studies*, edited by S. Jasanoff et al. London: Sage, 321-42.

Ashton, W. 2009. Ouija. *Fortean Times*, 254, October, 74.

Atack, J. 1990. *A Piece of Blue Sky: Scientology, Dianetics and L. Ron Hubbard Exposed*. New York: Carol Publishing Group.

Atkinson, P. and Housley, M. 2003. *Interactionism*. London: Sage.

Bainbridge, W.S. 1986. *Dimensions of Science Fiction*. Cambridge, MA: Harvard University Press.

Baker, C. 2000. Locating culture in action: membership categorisation in texts and talk, in *Culture and Text: Discourse and Methodology in Social Research and Cultural Studies*, edited by A. Lee and C. Poynton. London: Allen and Unwin, 99-113.

Bakhtin, M. 1981. *The Dialogic Imagination: Four Essays*, translated by C. Emerson and M. Holquist. Austin: University of Texas Press.

Barker, C. 2008. *Cultural Studies: Theory and Practice*. 3rd Edition. London: Sage.

Barker, M. 1984. *A Haunt of Fears: The Strange History of the British Horror Comics Campaign*. London: Pluto.

Barker, M. 1989. *Comics: Ideology, Power and the Critics*. Manchester: Manchester University Press.

Barker, M. 1993. Seeing how far you can see: on being a "fan" of *2000 AD*, in *Reading Audiences: Young People and the Media*, edited by D. Buckingham. Manchester: Manchester University Press, 159-83.

Barker, M. 1997. Taking the extreme case: understanding a fascist fan of Judge Dredd, in *Trash Aesthetics: Popular Culture and its Audience*, edited by D. Cartmell et al. London: Pluto, 14-30.

Barker, M. and Brooks, K. 1998. *Knowing Audiences:* Judge Dredd*, its Friends, Fans and Foes*. Luton: University of Luton Press.

Barkun, M. 2003. *A Culture of Conspiracy: Apocalyptic Visions in Contemporary America*. Berkeley: University of California Press.

Barnes, B. 1974. *Scientific Knowledge and Sociological Theory*. London: Routledge and Kegan Paul.

Barnes, B. 1977. *Interests and the Growth of Knowledge*. London: Routledge and Kegan Paul.

Barnes, B. and Bloor, D. 1982. Relativism, rationalism and the sociology of knowledge, in *Rationality and Relativism*, edited by M. Hollis and S. Lukes. Oxford: Basil Blackwell, 21-47.

Barnes, B., Bloor, D. and Henry, J. 1996. *Scientific Knowledge: A Sociological Analysis*. London: Athlone.

Barnes, B. and Edge, D. (eds) 1982. *Science in Context: Readings in the Sociology of Science*. Milton Keynes: Open University Press.

Barnes, B. and Shapin, S. (eds) 1979. *Natural Order: Historical Studies of Scientific Culture*. London: Sage.

Bartholomew, R. and Evans, H. 2005. The Martians are coming. *Fortean Times* 199, August, 42-47.

Basalla, G. 1976. Pop science: the depiction of science in popular culture, in *Science and its Public: The Changing Relationship*, edited by G. Holton and W.A. Blanpied. Dordrecht-Holland: D. Reidel, 261-78.

Baudrillard, J. 1998. *The Consumer Society: Myths and Structures*. London: Sage.

Bazerman, C. 1981. What written knowledge does: three examples of academic discourse. *Philosophy of the Social Sciences*, 11, 361-87.

Bazerman, C. 1987. Codifying the social scientific style: the APA *Publication Manual* as a behaviorist rhetoric, in *The Rhetoric of the Human Sciences: Language and Argument in Scholarship and Public Affairs*, edited by J.S.

Nelson, A. Megill and D.N. McCloskey. Madison, WI: University of Wisconsin Press, 125-44.

Bazerman, C. 1988. *Shaping Written Knowledge*. Madison, WI: University of Wisconsin Press.

Bazerman, C. 1997. Reporting the experiment: the changing account of scientific doings in the philosophical transactions of the Royal Society, 1665-1800, in *Landmark Essays on Rhetoric of Science: Case Studies*, edited by R.A. Harris. Mahwah, NJ: Lawrence Erlbaum Associates, 169-86.

Beck, U. 1992. *Risk Society: Towards a New Modernity*, translated by M. Ritter. London: Sage.

Beer, G. and Martins, H. 1990. Introduction. *History of the Human Sciences*, 3, 163-75.

Begg, P., Fido, M. and Skinner, K. 1991. *The Jack the Ripper A to Z*. London: Headline.

Bell, D. 1976. *The Cultural Contradictions of Capitalism*. London: Heinemann.

Bell, D. 2006. *Science, Technology and Culture*. Maidenhead: Open University Press.

Bell, D. and Bennion-Nixon, L.-J. 2001. The popular culture of conspiracy / the conspiracy of popular culture, in *The Age of Anxiety: Conspiracy Theory and the Human Sciences*, edited by J. Parish and M. Parker. Oxford: Blackwell, 133-52.

Bennett, T. and Woollacott, J. 1987. *Bond and Beyond: The Political Career of a Popular Hero*. Basingstoke: Macmillan.

Berger, A.I. 1989. Towards a science of the nuclear mind: science-fiction origins of Dianetics. *Science Fiction Studies*, 16, 123-44.

Berger, P. and Luckmann, T. 1967. *The Social Construction of Reality: A Treatise in the Sociology of Knowledge*. Harmondsworth: Penguin.

Bernard Shaw, G. 1932[1911]. *Plays. Selections*. London: Constable.

Besecke, K. 2001. Speaking of meaning in modernity: reflexive spirituality as a cultural resource. *Sociology of Religion*, 62(3), 365-81.

Billig, M. 1991. *Ideology and Opinions: Studies in Rhetorical Psychology*. London: Sage.

Billig, M. 1996. *Arguing and Thinking: A Rhetorical Approach to Social Psychology*. 2nd Edition. Cambridge: Cambridge University Press.

Billig, M. 1997. From codes to utterances: cultural studies, discourse and psychology, in *Cultural Studies in Question*, edited by M. Ferguson and P. Golding. London: Sage, 205-26.

Billig, M. 1999a. Whose terms? Whose ordinariness? Rhetoric and ideology in conversation analysis. *Discourse and Society*, 10(4), 543-58.

Billig, M. 1999b. Conversation analysis and claims of naivety. *Discourse and Society*, 10(4), 572-76.

Billig, M. 2004. Methodology and scholarship in understanding ideological explanation, in *Social Research Methods: A Reader*, edited by C. Seale. London: Routledge, 13-18.

Billig, M. 2005. *Laughter and Ridicule: Towards a Social Critique of Humour*. London: Sage.

Billig, M. 2008a. The language of critical discourse analysis: the case of nominalization. *Discourse and Society*, 19(6), 783-800.

Billig, M. 2008b. Nominalizing and de-nominalizing: a reply. *Discourse and Society*, 19(6), 829-41.

Billig, M., Condor, S., Edwards, D., Gane, M., Middleton, D. and Radley, A. 1988. *Ideological Dilemmas: A Social Psychology of Everyday Thinking*. London: Sage.

Bittner, E. 1974. The concept of organization, in *Ethnomethodology: Selected Readings*, edited by R. Turner. Harmondsworth: Penguin, 69-81.

Blaikie, N. 2007. *Approaches to Social Enquiry: Advancing Knowledge*. 2nd Edition. Cambridge: Polity.

Blasi, A.J. 2009. A market theory of religion. *Social Compass*, 56(2), 263-72.

Bloor, D. 1976. *Knowledge and Social Imagery*. London: Routledge and Kegan Paul.

Bloor, D. 1992. Left and right Wittgensteinians, in *Science as Practice and Culture*, edited by A. Pickering. Chicago: University of Chicago Press, 266-82.

Blythe, H. and Sweet, C. 1983. Superhero: the six step progression, in *The Hero in Transition*, edited by R.B. Browne and M.W. Fishwick. Bowling Green, OH: Bowling Green University Press, 180-7.

Boden, D. 1994. *The Business of Talk: Organizations in Action*. Cambridge: Polity.

Bongco, M. 2000. *Reading Comics: Language, Culture, and the Concept of the Superhero in Comic Books*. New York: Garland.

Bowles, S. and Gintis, H. 1976. *Schooling in Capitalist America: Educational Reform and the Contradictions of Economic Life*. London: Routledge and Kegan Paul.

Broks, P. 2006. *Understanding Popular Science*. Maidenhead: Open University Press.

Brooker, W. 2000. *Batman Unmasked: Analyzing a Cultural Icon*. London: Continuum.

Brookesmith, P. 2008. When aliens go to college. *Fortean Times*, 238, July, 48-53.

Brookesmith, P. and Irving, R. 2009a. Hoax! *Fortean Times*, 250, Special, 40-45.

Brookesmith, P. and Irving, R. 2009b. Hoax! Part 2. *Fortean Times*, 251, July, 48-53.

Brookesmith, P. and Irving, R. 2009c. Hoax! Part 3. *Fortean Times*, 252, August, 36-41.

Brossard, D. 2009. Media, scientific journals and science communication: examining the construction of scientific controversies. *Public Understanding of Science*, 18(3): 258-74.

Brown, J.A. 1997. Comic book fandom and cultural capital. *Journal of Popular Culture*, 30(4), 13-31.

Brown, J.A. 2001. *Black Superheroes,* Milestone Comics *and their Fans*. Jackson: University Press of Mississippi.

Brown, R.H. 1994. Logics of discovery as narratives of conversion: rhetorics of invention in ethnography, philosophy, and astronomy. *Philosophy and Rhetoric*, 27, 1-34.

Burr, V. 2003. *Social Constructionism*. 2nd Edition. London: Routledge.

Byford, J. 2002. Anchoring and objectifying "neocortical warfare": re-presentation of a biological metaphor in Serbian conspiracy literature. *Papers on Social Representations* [Online], 11, 3.1-3.14. Available at: http://ww.psych.lse.ac.uk/psr/PSR2002/11_3Byfor.pdf [accessed: 4 March 2010].

Byrne, J. 1986. *The Man of Steel* 1, June. New York: DC Comics.

Bytwerk, R.L. 1979. The SST controversy: a case study of the rhetoric of technology. *Central States Speech Journal*, 30, 187-98.

Cambrosio, A., Keating, P. and MacKenzie, M. 1990. Scientific practice in the courtroom: the construction of sociotechnical identities in a biotechnology patent dispute. *Social Problems*, 37(3), 275-91.

Cameron, D. and Fraser, E. 1987. *Lust to Kill: A Feminist Investigation of Sexual Murder*. New York: New York University Press.

Cameron, I. and Edge, D. 1979. *Scientific Images and their Social Uses: An Introduction to the Concept of Scientism*. London: Butterworths.

Campbell, C. 1972. The cult, the cultic milieu and secularisation, in *A Sociological Yearbook of Religion in Britain 5*, edited by M. Hill. London: SCM, 119-36.

Campbell, C. 1987. *The Romantic Ethic and the Spirit of Modern Consumerism*. Oxford: Blackwell.

Campbell, J.A. 1997. Charles Darwin: rhetorician of science, in *Landmark Essays on Rhetoric of Science: Case Studies*, edited by R.A. Harris. Mahwah, NJ: Lawrence Erlbaum Associates, 3-17.

Campbell, P.N. 1975. The *personae* of scientific discourse. *Quarterly Journal of Speech*, 61, 391-405.

Campbell, R.A. 2001. The truth will set you free: towards the religious study of science. *Journal of Contemporary Religion*, 16(1), 29-43.

Cantril, H. 2005. *The Invasion from Mars: A Study in the Psychology of Panic*. Princeton, NJ: Princeton University Press.

Catron, M. 1996. Superman creator Jerry Siegel dies at 81. *The Comics Journal*, 184, February, 25-39.

Chabon, M. 2000. *The Amazing Adventures of Kavalier and Clay*. London: Fourth Estate.

Chaney, D. 2002. *Cultural Change and Everyday Life*. Basingstoke: Palgrave.

Children and Young Persons (Harmful Publications) Bill 1955, London: HMSO.

Church of Scientology International, 1998a. *Scientology: Theology and Practice of a Contemporary Religion*. Los Angeles: Bridge.

Church of Scientology International, 1998b. *What is Scientology?* Los Angeles: Bridge.

Citizen's Commission on Human Rights (CCHR), 2000. *Documenting Psychiatry: A Human Rights Abuse and Global Failure*. London: CCHR.
Clarke, S. 2002. Conspiracy theory and conspiracy theorizing. *Philosophy of the Social Sciences*, 32(2), 131-50.
Clute, J. and Nicholls, P. (eds) 1993. *The Encyclopaedia of Science Fiction*. London: Orbit.
Clute, J. and Grant, J. 1997. Introduction, in *Encyclopedia of Fantasy*, edited by J. Clute and J. Grant. London: Orbit, vii-x.
Cole, S.A. 2006. Witnessing creation. *Social Studies of Science*, 36(6), 855-60.
Coleman, L. 2004. Costume drama. *Fortean Times*, 185, July, 58-59.
Coleman, S. and Carlin, L. (eds.) 2004. *The Cultures of Creationism: Anti-Evolutionism in English-Speaking Countries*. Aldershot: Ashgate.
Collins. 1994. *Collins English Dictionary*. 3rd Updated Edition. Glasgow: HarperCollins.
Collins, H.M. 1983. An empirical relativist programme in the sociology of scientific knowledge, in *Science Observed: Perspectives on the Social Study of Science*, edited by K. Knorr-Cetina and M. Mulkay. London: Sage, 85-113.
Collins, H.M. 1985. *Changing Order: Replication and Induction in Scientific Practice*. London: Sage.
Collins, H.M. 1987. Certainty and the public understanding of science: science on television. *Social Studies of Science*, 17, 689-713.
Collins, H.M. 1988. Public experiments and displays of virtuosity: the core set revisited. *Social Studies of Science*, 18, 725-48.
Collins, H.M. 2004. *Gravity's Shadow: The Search for Gravitational Waves*. Chicago: University of Chicago Press.
Collins, H.M. and Evans, R. 2002. The third wave of science studies. *Social Studies of Science*, 32(2), 235-96.
Collins, H.M. and Pinch, T. 1979. The construction of the paranormal: nothing unscientific is happening, in *On the Margins of Science: The Social Construction of Rejected Knowledge*, edited by R. Wallis. Keele: Keele University Press, 237-70.
Collins, H.M. and Pinch, T. 1982. *Frames of Meaning: The Social Construction of Extraordinary Science*. Henley-on-Thames: Routledge and Kegan Paul.
Collins, H.M. and Pinch, T. 1993. *The Golem: What Everyone Should Know about Science*. Cambridge: Cambridge University Press.
Collins, H.M. and Yearley, S. 1992. Epistemological chicken, in *Science as Practice and Culture*, edited by A. Pickering. Chicago: University of Chicago Press, 301-26.
Coogan, P. 2006. *Superhero: The Secret Origin of a Genre*. Austin: MonkeyBrain Books.
Cooter, R. and Pumfrey, S. 1994. Separate spheres and public places: reflections on the history of science popularization and science in popular culture. *History of Science*, 32, 237-67.

Cremo, M.A. 1998. *Forbidden Archeology's Impact: How a Controversial New Book Shocked the Scientific Community and Became an Underground Classic.* Los Angeles: Bhaktivedanta.

Crook, S., Pakulski J. and Waters, M. 1992. *Postmodernization: Change in Advanced Society*. London: Sage.

CSI 2010. *Committee for Skeptical Inquiry homepage*. [Online]. Available at: http://www.csicop.org [accessed: 8 April 2010].

Cuff, E.C., Sharrock, W.W. and Francis, D.W. 2006. *Perspectives in Sociology*. 5th Edition. Abingdon: Routledge.

Curran, J. and Morley, D. 2006. Editors' introduction, in *Media and Cultural Theory*, edited by J. Curran and D. Morley. Abingdon: Routledge, 1-13.

Curtis, L.P. 2001. *Jack the Ripper and the London Press*. New Haven: Yale University Press.

Davie, G. 2007. *The Sociology of Religion*. London: Sage.

Davis, E. 1998. *TechGnosis: Myth, Magic and Mysticism in the Age of Information.* New York: Harmony.

Dawkins, R. 1998. *Unweaving the Rainbow: Science, Delusion and the Appetite for Wonder*. London: Allen Lane.

Dawkins, R. 2006. *The Selfish Gene 30th Anniversary Edition*. Oxford: Oxford University Press.

Dawson, L. 1998a. Anti-modernism, modernism and postmodernism: struggling with the cultural significance of new religious movements. *Sociology of Religion*, 59, 131-56.

Dawson, L. 1998b. The cultural significance of new religious movements and globalization: a theoretical prolegomenon. *Journal for the Scientific Study of Religion*, 37, 580-95.

Day, D. 1998. Being ascribed, and resisting, membership of an ethnic group, in *Identities in Talk*, edited by C. Antaki and S. Widdicombe. London: Sage, 151-70.

Dean, J. 1998. *Aliens in America: Conspiracy Cultures from Outerspace to Cyberspace*. Ithaca: Cornell University Press.

Dear, P. 1991. Narratives, anecdotes, and experiments: turning experience into science in the seventeenth century, in *The Literary Structure of Scientific Argument: Historical Studies*, edited by P. Dear. Philadelphia: University of Pennsylvania Press, 135-63.

Delamont, S. 2009. Neopagan narratives: knowledge claims and other world "realities". *Sociological Research Online* [Online], 14(5). Available at: http://www.socresonline.org.uk/14/5/18.html [accessed: 24 March 2010].

Delanty, G. and Strydom, P. (eds) 2003. *Philosophies of Social Science*. Maidenhead: Open University Press.

Denzler, B. 2003. *The Lure of the Edge: Scientific Passions, Religious Beliefs and the Pursuit of UFOs*. Berkeley: University of California Press.

Derksen, M. 1997. Are we not experimenting then? The rhetorical demarcation of psychology and common sense. *Theory and Psychology*, 7(4), 435-56.

Devereux, P. 1991. *Earth Memory: The Holistic Earth Mysteries Approach to Decoding Ancient Sacred Sites*. London: Quantum.

Devereux, P. 2007. Where the leylines led. *Fortean Times*, 221, April, 30-36.

Dingwall, R. 1976. Accomplishing profession. *Sociological Review*, 24, 331-49.

Dittmer, J. 2007. The tyranny of the serial: popular geopolitics, the nation and comic book discourse. *Antipode*, 39(2), 247-68.

Dittmer, J. Forthcoming. Captain Britain and the narration of nation. *Geographical Review*.

Douglas, A. 1974. *The Tarot: The Origins, Meaning and Uses of the Cards*. Harmondsworth: Penguin.

Douglas, M. 1992. Risk and blame, in *Risk and Blame: Essays in Cultural Theory*. London: Routledge, 3-21.

Dunbar, R. 1995. *The Trouble with Science*. London: Faber.

Durant, D. 2008. Accounting for expertise: Wynne and the autonomy of the lay public actor. *Public Understanding of Science*, 17(1), 5-20.

Durant, J. 1993. What is scientific literacy?, in *Science and Culture in Europe*, edited by J. Durant and J. Gregory. London: Science Museum, 129-38.

Durkheim, E. 1964. The dualism of human nature and its social conditions, in *Emile Durkheim: Essays on Sociology and Philosophy*, edited by K. Wolff. New York: Harper Torchbooks, 325-40.

Eagleton, T. 1991. *Ideology: An Introduction*. London: Verso.

Eco, U. 1979. The myth of Superman, in *The Role of the Reader: Explorations in the Semiotics of Texts*. London: Indiana University Press, 107-24.

Edley, N. 1993. Prince Charles – our flexible friend: accounting for variations in constructions of identity. *Text*, 13(3), 397-422.

Edmond, G. and Mercer, D. 2006. Anti-social epistemologies. *Social Studies of Science*, 36(6), 843-53.

Edmondson, R. 1984. *Rhetoric in Sociology*. London: Macmillan.

Edwards, D. and Potter, J. 1992. *Discursive Psychology*. London: Sage.

Eglin, P. and Hester, S. 1999. Moral order and the Montreal massacre: a story of Membership Categorization Analysis, in *Media Studies: Ethnomethodological Approaches*, edited by P.L. Jalbert. Lanham: University Press of America, 195-230.

Einsedel, E.F. 2000. Understanding "publics" in the public understanding of science, in *Between Understanding and Trust: The Public, Science and Technology*, edited by M. Dierkes and C. von Grote. Amsterdam: Harwood, 205-15.

Eldredge, N. 2000. *The Triumph of Evolution and the Failure of Creationism*. New York: W.H. Freeman and Co.

Eldridge, J.E.T. 1971. *Max Weber: The Interpretation of Social Reality*. London: Nelson.

Erickson, M. 2005. *Science, Culture and Society: Understanding Science in the Twenty-First Century*. Cambridge: Polity.

Escobar, A. 2000. Welcome to Cyberia: notes on the anthropology of cyberculture, in *The Cybercultures Reader*, edited by D. Bell and B. Kennedy. London: Routledge, 56-76.

Evans, G. and Durant, J. 1989. Understanding of science in Britain and the USA, in *British Social Attitudes: Special International Report*, edited by R. Jowell, S. Witherspoon and L. Brook. Aldershot: Gower, 105-19.

Evans, H. and Stacy, D. (eds) 1997. *UFOs 1947-1997. From Arnold to the Abductees: Fifty Years of Flying Saucers*. London: John Brown.

Evans, M.S. 2009. Defining the public, defining sociology: hybrid science-public relations and boundary-work in early American sociology. *Public Understanding of Science*, 18(1), 5-22.

Fahnestock, J. 1986. Accommodating science. *Written Communication*, 3, 275-96.

Fahnestock, J. 1997. Arguing in different forums: the Bering Crossover controversy, in *Landmark Essays on Rhetoric of Science: Case Studies*, edited by R.A. Harris. Mahwah, NJ: Lawrence Erlbaum Associates, 53-67.

Fairclough, N. 2008a. The language of critical discourse analysis: reply to Michael Billig. *Discourse and Society*, 19(6), 811-19.

Fairclough, N. 2008b. A brief response to Billig. *Discourse and Society*, 19(6), 843-44.

Ferguson, M. and Golding, P. (eds) 1997. *Cultural Studies in Question*. London: Sage.

Feyerabend, P. 1993. *Against Method*. 3rd Edition. London: Verso.

Fisher, W.R. 1994. Narrative rationality and the logic of scientific discourse. *Argumentation*, 8, 21-32.

Fiske, J. 1998. *Understanding Popular Culture*. London: Routledge.

Flanagan, K. 2007. Visual spirituality: an eye for religion, in *A Sociology of Spirituality*, edited by K. Flanagan and P.C. Jupp. Aldershot: Ashgate, 219-49.

Flory, R.W. and Miller, D.E. 2007. The embodied spirituality of the post-boomer generations, in *A Sociology of Spirituality*, edited by K. Flanagan and P.C. Jupp. Aldershot: Ashgate, 201-18.

Foltz, T.G. 2006. Drumming and re-enchantment: creating spiritual community, in *Popular Spiritualities: The Politics of Contemporary Enchantment*, edited by L. Hume and K. McPhillips. Aldershot: Ashgate, 131-43.

Fort, C.H. 1995. *The Book of the Damned*. Revised Edition. London: John Brown.

Fort, C.H. 1996. *New Lands*. Revised Edition. London: John Brown.

Fort, C.H. 1997. *Lo!* Revised Edition. London: John Brown.

Fort, C.H. 1998. *Wild Talents*. Revised edition. London: John Brown.

Foucault, M. 1970. *The Order of Things: An Archaeology of the Human Sciences*. London: Tavistock.

Foucault, M. 2001. *Madness and Civilization: A History of Insanity in the Age of Reason*, translated by R. Howard. London: Routledge.

Frances, D. and Hester, S. 2004. *An Invitation to Ethnomethodology: Language, Society and Interaction*. London: Sage.

Frayling, C. 1986. The house that Jack built: some stereotypes of the rapist in the history of popular culture, in *Rape*, edited by S. Tomaselli and R. Porter. Oxford: Basil Blackwell, 174-215.

Fulcher, J. and Scott, J. 2007. *Sociology*. 3rd Edition. Oxford: Oxford University Press.

Fuller, S. 1993. *Philosophy, Rhetoric and the End of Knowledge: The Coming of Science and Technology Studies*. Madison, WI: University of Wisconsin Press.

Fuller, S. 1997. "Rhetoric of science": double the trouble?, in *Rhetorical Hermeneutics: Invention and Interpretation in the Age of Science*, edited by A.G. Gross and W.M. Keith. Albany, NY: State University of New York Press, 279-98.

Fuller, S. 2000. *The Governance of Science: Ideology and the Future of the Open Society*. Buckingham: Open University Press.

Fuller, S. 2006. A step toward the legalization of science studies. *Social Studies of Science*, 36(6), 827-34.

Fuller, S. 2007. *Science Vs Religion? Intelligent Design and the Problem of Evolution*. Cambridge: Polity.

Gabriel, Y. and Lang, T. 1995. *The Unmanageable Consumer: Contemporary Consumption and its Fragmentations*. London: Sage.

Gane, N. 2004. *Max Weber and Postmodern Theory: Rationalization Versus Re-enchantment*. Basingstoke: Palgrave Macmillan.

Gaonkar, D.P. 1997. The idea of rhetoric in the rhetoric of science, in *Rhetorical Hermeneutics: Invention and Interpretation in the Age of Science*, edited by A.G. Gross and W.M. Keith. Albany, NY: State University of New York Press, 25-85.

Gardner, M. 1957. Dianetics, in *Fads and Fallacies in the Name of Science*. 2nd Edition [Online]. Available at http://www.cs.cmu.edu/~dst/Library/Shelf/gardner/ [accessed: 4 March 2010].

Garfinkel, H. 1967. *Studies in Ethnomethodology*. Englewood Cliffs: Prentice-Hall.

Garfinkel, H. 1974. The origins of the term "ethnomethodology", in *Ethnomethodology: Selected Readings*, edited by R. Turner. Harmondsworth: Penguin, 15-18.

Garfinkel, H. 1991. Respecification: evidence for locally produced, naturally accountable phenomena of order, logic, reason, meaning, method, etc. in and as of the essential haecceity of immortal ordinary society (I) – an announcement of studies, in *Ethnomethodology and the Human Sciences*, edited by G. Button. Cambridge: Cambridge University Press, 10-19.

Garfinkel, H. 2002. *Ethnomethodology's Program: Working Out Durkheim's Aphorism*. Lanham, MD: Rowman and Littlefield.

Garfinkel, H., Lynch, M. and Livingstone, E. 1981. The work of a discovering science construed with materials from the optically discovered pulsar. *Philosophy of the Social Sciences*, 11, 131-58.

Garnham, N. 1997. Political economy and the practice of cultural studies, in *Cultural Studies in Question*, edited by M. Ferguson and P. Golding. London: Sage, 56-73.

Garnham, N. 1998. Political economy and cultural studies: reconciliation or divorce?, in *Cultural Theory and Popular Culture: A Reader*. 2nd Edition, edited by J. Storey. Harlow: Longman, 600-12.

Gergen, K. 2009. *An Invitation to Social Construction*. 2nd Edition. London: Sage.

Giddens, A. 1971. *Capitalism and Modern Social Theory: An Analysis of the Writings of Marx, Durkheim and Max Weber*. Cambridge: Cambridge University Press.

Giddens, A. 1984. *The Constitution of Society: Outline of the Theory of Structuration*. Cambridge: Polity.

Giddens, A. 1991. *Modernity and Self-identity: Self and Society in the Late Modern Age*. Cambridge: Polity.

Gieryn, T.F. 1983. Boundary-work and the demarcation of science from non-science: strains and interests in professional ideologies of scientists. *American Sociological Review*, 48, 781-95.

Gieryn, T.F. 1995. Boundaries of science, in *Handbook of Science and Technology Studies*, edited by S. Jasanoff et al. London: Sage, 393-443.

Gieryn, T.F. 1999. *Cultural Boundaries of Science: Credibility on the Line*. Chicago: University of Chicago Press.

Gieryn, T.F. and Figert, A.E. 1990. Ingredients for a theory of science in society: o-rings, ice-water, c-clamp, Richard Feynman and the press, in *Theories of Science in Society*, edited by S.E. Cozzens and T.F. Gieryn. Bloomington: Indiana University Press, 67-97.

Gilbert, G.N. 1977. Referencing as persuasion. *Social Studies of Science*, 7, 113-22.

Gilbert, G.N. and Mulkay, M. 1984. *Opening Pandora's Box: A Sociological Analysis of Scientists' Discourse*. Cambridge: Cambridge University Press.

Goffman, E. 1990. *The Presentation of Self in Everyday Life*. London: Penguin.

Goldacre, B. 2008. *Bad Science*. London: Fourth Estate.

Gordon, D. 2009. Investigating alien abduction. *Fortean Times*, 251, July, 75.

Greenberg, J.M. 2004. Creating the "Pillars": multiple meanings of a Hubble image. *Public Understanding of Science*, 13, 83-95.

Gregory, J. and Miller, S. 1998. *Science in Public: Communication, Culture and Credibility*. New York: Plenum Press.

Gresh, L. and Weinberg, R. 2002. *The Science of Superheroes*. Hoboken, NJ: John Wiley and Sons.

Gribbin, J. 2000. *Stardust*. Harmondsworth: Penguin.

Griffin, D.R. (ed.) 1988. *The Reenchantment of Science: Postmodern Proposals*. New York: State University of New York Press.

Groh, D. 1987. The temptation of conspiracy theory, or: why do bad things happen to good people? Part I: preliminary draft of a theory of conspiracy theories, in *Changing Conceptions of Conspiracy*, edited by C.F. Graumann and S. Moscovici. New York: Springer-Verlag, 1-13.

Gross, A.G. 1991. Does rhetoric of science matter? The case of the floppy-eared rabbits. *College English*, 53, 933-43.

Gross, A.G. 1994a. Is a rhetoric of science policy possible? *Social Epistemology*, 8, 273-80.

Gross, A.G. 1994b. The roles of rhetoric in the public understanding of science. *Public Understanding of Science*, 3, 3-23.

Gross, A.G. 1995. Renewing Aristotelian theory: the cold fusion theory as a test case. *Quarterly Journal of Speech*, 81, 48-62.

Gross, A.G. 1996. *The Rhetoric of Science.* 2nd Edition. Cambridge, MA: Harvard University Press.

Gross, P.R. and Levitt, N. 1994. *Higher Superstition: The Academic Left and Its Quarrels with Science*. Baltimore: Johns Hopkins University Press.

Gross, P.R., Levitt, N. and Lewis, M. (eds) 1997. *The Flight from Reason and Science.* New York: New York Academy of Sciences.

Grossberg, L. 1998. Cultural studies vs political economy: is anybody else bored with this debate?, in *Cultural Theory and Popular Culture: A Reader*. 2nd Edition, edited by J. Storey. Harlow: Longman, 613-24.

Gruenwald, M. 1986. Scientific method. *Official Handbook of the Marvel Universe*, 2(3), February. New York: Marvel Comics, inside front cover.

Grundmann, R. and Cavaillé, J.-P. 2000. Simplicity in science and its publics. *Science as Culture*, 9(3), 353-89.

Habermas, J. 1971. Technology and science as "ideology", in *Toward a Rational Society: Student Protest, Science and Politics*, translated by J. Shapiro. London: Heinemann, 81-127.

Habermas, J. 1973. What does a crisis mean today? Legitimation problems in late capitalism. *Social Research*, 40, 643-67.

Habermas, J. 1984. *The Theory of Communicative Action, Volume 1: Reason and the Rationalization of Society*, translated by T. McCarthy. London: Heinemann.

Habermas, J. 1987. *The Theory of Communicative Action Volume 2: Lifeworld and System, a Critique of Functionalist Reason*, translated by T. McCarthy. Cambridge: Polity.

Habermas, J. 1989. *The Structural Transformation of the Public Sphere: An Inquiry into a Category of Bourgeois Society*, translated by T. Burger and F. Lawrence. Cambridge: Polity.

Hajdu, D. 2008. *The Ten-Cent Plague: The Great Comic Book Scare and How it Changed America.* New York: Picador.

Hall, S. 1980. Encoding/decoding, in *Culture, Media, Language: Working Papers in Cultural Studies 1972-79*, edited by S. Hall et al. London: Hutchinson, 128-38.

Hall, S., Hobson, D., Lowe, A. and Willis, P. (eds) 1980. *Culture, Media, Language: Working Papers in Cultural Studies 1972-79*. London: Hutchinson.

Hall, S. and Jefferson, T. (eds) 1976. *Resistance Through Rituals: Youth Subcultures in Post-War Britain*. London: Hutchinson.

Halliday, M.A.K. 1998. Things and relations: regrammaticising experience as technical knowledge, in *Reading Science: Critical and Functional Perspectives on Discourses of Science*, edited by J.R. Martin and R. Veel. London: Routledge, 185-235.

Halloran, S.M. 1997. The birth of molecular biology: an essay in the rhetorical criticism of scientific discourse, in *Landmark Essays on Rhetoric of Science: Case Studies*, edited by R.A. Harris. Mahwah, NJ: Lawrence Erlbaum Associates, 39-50.

Hand, S. and Velody, I. 1997. Introduction. *History of the Human Sciences*, 10, 1-8.

Hanegraaff, W.L. 1999. New Age spiritualities as secular religion: a historian's perspective. *Social Compass*, 46(2), 145-60.

Haraway, D. 1991. *Simians, Cyborgs and Women: the Reinvention of Nature*. London: Free Association.

Harding, P. and Jenkins, R. 1989. *The Myth of the Hidden Economy: Towards a New Understanding of Informal Economic Activity*. Milton Keynes: Open University Press.

Harding, S. 1991. *Whose Science? Whose Knowledge? Thinking from Women's Lives*. Ithaca, NY: Cornell University Press.

Harpur, P. 2009. Capturing the secret fire. *Fortean Times*, 246, March, 60-61.

Harrington, A. 1996. *Reenchanted Science: Holism in German Culture from Wilhelm II to Hitler*. Princeton, NJ: Princeton University Press.

Harris, R.A. (ed.) 1997. *Landmark Essays on Rhetoric of Science: Case Studies*. Mahwah, NJ: Lawrence Erlbaum Associates.

Hatfield, C. 1998. On Kirby's "unexpected constants": the problem of Marvel continuity in *The Eternals*. *The Jack Kirby Collector*, 18, January, 54-57.

Hatfield, C. 2005. *Alternative Comics: An Emerging Literature*. Jackson: University Press of Mississippi.

Hawking, S. 1988. *A Brief History of Time From the Big Bang to Black Holes*. London: Bantam.

Haynes, R.D. 1994. *From Faust to Strangelove: Representations of the Scientist in Western Literature*. Baltimore: Johns Hopkins University Press.

Haynes, R.D. 2003. From alchemy to artificial intelligence: stereotypes of the scientist in western literature. *Public Understanding of Science*, 12, 243-53.

Hebdige, D. 1979. *Subculture: The Meaning of Style*. London: Methuen.

Heelas, P. 1991. Reforming the self: enterprise and the characters of Thatcherism, in *Enterprise Culture*, edited by R. Keat and N. Abercrombie. London: Routledge, 72-90.

Heelas, P. 1996. *The New Age Movement: The Celebration of the Self and the Sacralization of Modernity*. Oxford: Blackwell.

Hennis, W. 1983. Max Weber's "central question". *Economy and Society*, 12(2), 135-80.

Hermes, J. 1995. *Reading Women's Magazines: Analysis of Everyday Media Use*. Cambridge: Polity.

Hess, D.J. 1993. *Science in the New Age: The Paranormal, its Defenders and Debunkers, and American Culture.* Madison, WI: University of Wisconsin Press.

Hess, D. 1997. *Science Studies: An Advanced Introduction.* New York: New York University Press.

Hester, S. 1998. Describing "deviance" in school: recognizably educational psychological problems, in *Identities in Talk*, edited by C. Antaki and S. Widdicombe. London: Sage, 133-50.

Hester, S. and Eglin, P. (eds) 1997a. *Culture in Action: Studies in Membership Categorization Analysis.* Washington DC: International Institute for Ethnomethodology and Conversation Analysis and University Press of Washington.

Hester, S. and Eglin, P. 1997b. Membership Categorization Analysis: an introduction, in *Culture in Action: Studies in Membership Categorization Analysis*, edited by S. Hester and P. Eglin. Washington DC: International Institute for Ethnomethodology and Conversation Analysis and University Press of Washington, 1-23.

Hester, S. and Eglin, P. 1997c. The reflexive constitution of category, predicate and context in two settings, in *Culture in Action: Studies in Membership Categorization Analysis*, edited by S. Hester and P. Eglin. Washington DC: International Institute for Ethnomethodology and Conversation Analysis and University Press of Washington, 25-48.

Hierophant's Apprentice. 2007. The *Fortean Times* random dictionary of the damned no. 17: conspiracy theory. *Fortean Times*, 223, June, 51-53.

Hilgartner, S. 1990. The dominant view of popularization: conceptual problems, political uses. *Social Studies of Science*, 20: 519-39.

Hills, M. 2002. *Fan Cultures*. London: Routledge.

Hodge, R. and Kress, G. 1993. *Language as Ideology*. 2nd Edition. London: Routledge.

Hofstadter, R. 1965. *The Paranoid Style in American Politics and Other Essays.* New York: Knopf.

Holman, B. 2008. Goodbye, Zeta Reticuli. *Fortean Times*, 242, November, 50-52.

Holton, G. 1992. How to think about the "anti-science" phenomenon. *Public Understanding of Science*, 1, 103-28.

Holton, G. 1993. *Science and Anti-science.* Cambridge, MA: Harvard University Press.

House of Lords. 2000. *Science and Society, Report of House of Lords Select Committee on Science and Technology, Chair Patrick Jenkin* (HL Paper 38), London: HMSO.

Hubbard, L. Ron. 1987. *Dianetics: The Modern Science of Mental Health.* Los Angeles: Bridge.

Hubbard, L. Ron. 1996. *The Rediscovery of the Human Soul.* Los Angeles: L. Ron Hubbard Library.

Hume, L. and McPhillips, K. (eds) 2006. *Popular Spiritualities: The Politics of Contemporary Enchantment*. Aldershot: Ashgate.

Hüppauf, B. and Weingart, P. 2008. Images in and of science, in *Science Images and Popular Images of the Sciences*, edited by B. Hüppauf and P. Weingart. London: Routledge, 3-31.

Ingram-Waters, M.C. 2009. Public fiction as knowledge production: the case of the Raëlians' cloning claims. *Public Understanding of Science*, 18(3), 292-308.

Irwin, A. 1995. *Citizen Science: A Study of People, Expertise and Sustainable Development*. London: Routledge.

Irwin, A. and Michael, M. 2003. *Science, Social Theory and Public Knowledge*. Buckingham: Open University Press.

Irwin, A. and Wynne, B. (eds) 1996. *Misunderstanding Science? The Public Reconstruction of Science and Technology*. Cambridge: Cambridge University Press.

Jancovich, M. 1996. *Rational Fears: American Horror in the 1950s*. Manchester: Manchester University Press.

Jasinski, J. 1997. Instrumentalism, contextualism and interpretation in rhetorical criticism, in *Rhetorical Hermeneutics: Invention and Interpretation in the Age of Science*, edited by A.G. Gross and W.M. Keith. Albany, NY: State University of New York Press, 195-224.

Jayyusi, L. 1984. *Categorization and the Moral Order*. London: Routledge and Kegan Paul.

Jayyusi, L. 1991. Values and moral judgement: communicative praxis as moral order, in *Ethnomethodology and the Human Sciences*, edited by G. Button. Cambridge: Cambridge University Press, 227-51.

Jenkins, H. 1992. *Textual Poachers: Television Fans and Participatory Culture*. London: Routledge.

Jenson, J. 1992. Fandom as pathology: the consequences of characterization, in *The Adoring Audience: Fan Culture and Popular Media*, edited by L. Lewis. London: Routledge, 9-29.

Johnson, T., Dandeker C. and Ashworth C. 1984. *The Structure of Social Theory: Dilemmas and Strategies*. Basingstoke: Macmillan.

Jones, G. 2006. *Men of Tomorrow: Geeks, Gangsters and the Birth of the Comic Book*. London: Arrow.

Jones, R.A. 1997. The boffin: a stereotype of scientists in post-war British films (1945-1970). *Public Understanding of Science*, 6(1), 31-48.

Jones, S. Forthcoming. *Porn of the Dead*: necrophilia, feminism, and gendering the undead, in *An Interdisciplinary Collection on the Zombie*, edited by C. Moreman and C. Rushton. Jefferson, NC: McFarland.

Jörg, D. 2003. The good, the bad and the ugly – Dr. Moreau goes to Hollywood. *Public Understanding of Science*, 12, 297-305.

Kaplan, A. 2008. *From Krakow to Krypton: Jews and Comic Books*. Philadelphia: Jewish Publication Society.

Kent, S.A. 1996. Scientology's relationship with eastern religious traditions. *Journal of Contemporary Religion*, 11(1), 21-36.

Kent, S.A. 1997. *Scientology – is this a religion?* 27th Deutscher Evangelischer Kirchentag, Leipzig, Germany, 20 June 1997.

Kent, S.A. 1999. The globalization of Scientology: influence, control and opposition in transnational markets. *Religion*, 29, 147-69.

Kerr, A., Cunningham-Burley, S. and Amos, A. 1997. The new genetics: professionals' discursive boundaries. *Sociological Review*, 45, 279-303.

Kidd, I. 2006. Who was Charles Fort? *Fortean Times*, 217, December, 54-55.

Kitcher, P. 1982. *Abusing Science: The Case Against Creationism*. Cambridge, MA: MIT Press.

Knight, P. 2000. *Conspiracy Culture: From Kennedy to* The X-Files. London: Routledge.

Knight, S. 1994. *Jack the Ripper: The Final Solution*. London: HarperCollins.

Knorr-Cetina, K. 1981. *The Manufacture of Knowledge: An Essay on the Constructivist and Contextual Nature of Science*. Oxford: Pergamon.

Korff, K. 2005. The making of Bigfoot. *Fortean Times*, 192, January, 34-39.

Kruglanski, A.W. 1987. Blame-placing schemata and attributional research, in *Changing Conceptions of Conspiracy*, edited by C.F. Graumann and S. Moscovici. New York: Springer-Verlag, 219-29.

Krupp, E.C. 1984. Observatories of the gods and other astronomical fantasies, in *In Search of Ancient Astronomies*, edited by E.C. Krupp. Harmondsworth: Penguin, 219-56.

Kuhn, T.S. 1970. *The Structure of Scientific Revolutions*. 2nd Edition. Chicago: University of Chicago Press.

Labinger, J. and Collins, H.M. (eds) 2001. *The One Culture? A Conversation About Science*. Chicago: University of Chicago Press.

LaFollette, M.C. 1983. *Creationism, Science and the Law: The Arkansas Case*. Cambridge, MA: MIT Press.

LaFollette, M.C. 1990. *Making Science Our Own: Public Images of Science 1910-1955*. Chicago: University of Chicago Press.

Laing, R.D. 1967. *The Politics of Experience and The Bird of Paradise*. Harmondsworth: Penguin.

Lakoff, G. and Johnson, M. 1980. *Metaphors We Live By*. Chicago: Chicago University Press.

Lambert, K. 2006. Fuller's folly, Kuhnian paradigms, and Intelligent Design. *Social Studies of Science*, 36(6), 835-42.

Lambourne, R., Shallis, M. and Shortland, M. 1990. *Close Encounters? Science and Science Fiction*. Bristol: Adam Hilger.

Lamont, P. 2010. Debunking and the psychology of error: a historical analysis of psychological matters. *Qualitative Research in Psychology*, 7(1), 34-44.

Larrain, J. 1979. *The Concept of Ideology*. London: Hutchinson.

Lash, S. and Urry, J. 1994. *Economies of Signs and Spaces*. London: Sage.

Lassman, P. and Velody, I. (eds) 1989. *Max Weber's "Science as a Vocation"*. London: Unwin Hyman.

Latour, B. 1987. *Science in Action: How to Follow Scientists and Engineers Through Society*. Cambridge, MA: Harvard University Press.

Latour, B. 1993. *We Have Never Been Modern*, translated by C. Porter. London: Harvester Wheatsheaf.

Latour, B. 2004. Why has critique run out of steam? From matters of fact to matters of concern. *Critical Inquiry*, 30, 225-48.

Latour, B. 2005. *Reassembling the Social: An Introduction to Actor-Network-Theory*. Oxford: Oxford University Press.

Latour, B. and Woolgar, S. 1986. *Laboratory Life: The Construction of Scientific Facts*. Princeton, NJ: Princeton University Press.

Lawrence, J.S. and Jewett, R. 2002. *The Myth of the American Superhero*. Grand Rapids: William B. Eerdmans.

Lee, R.L.M. 2003. The re-enchantment of the self: western spirituality, Asian materialism. *Journal of Contemporary Religion*, 18(3), 351-67.

Lee, R.L.M. 2007. Mortality and re-enchantment: conscious dying as individualized spirituality. *Journal of Contemporary Religion*, 22(2), 221-34.

Lee, R.L.M. 2008. Modernity, mortality and re-enchantment: the death taboo revisited. *Sociology*, 42(4), 745-59.

Lee, R.L.M. 2009. The re-enchantment of time: death and alternative temporality. *Time and Society*, 18(2/3), 387-408.

Lee, R.L.M. and Ackerman, S.E. 2002. *The Challenge of Religion After Modernity: Beyond Disenchantment*. Aldershot: Ashgate.

Lee, S. and Ditko, S. 1962. Spider-Man! *Amazing Fantasy* 15, September. New York: Atlas Magazines.

Lee, S. and Kirby, J. 1963a. The return of Doctor Doom! *Fantastic Four* 10, January. New York: Marvel Comics.

Lee, S. and Kirby, J. 1963b. X-Men. *The X-Men* 1, September. New York: Marvel Comics.

Lee, S. and Kirby, J. 1964. The fantastic origin of Doctor Doom! *Fantastic Four Annual* 2. New York: Marvel Comics.

Lent, J.A. (ed.) 1999. *Pulp Demons: International Dimensions of the Postwar Anti-Comics Campaign*. London: Associated University Presses.

Lepper, G. 2000. *Categories in Text and Talk: A Practical Introduction to Categorization Analysis*. London: Sage.

Lessl, T.M. 1985. Science and the sacred cosmos: the ideological rhetoric of Carl Sagan. *Quarterly Journal of Speech*, 71, 175-87.

Lessl, T.M. 1988. Heresy, orthodoxy, and the politics of science. *Quarterly Journal of Speech*, 74, 18-34.

Lessl, T.M. 1989. The priestly voice. *Quarterly Journal of Speech*, 75, 183-97.

Letcher, A. 2006. "There's bulldozers in the fairy garden": re-enchantment narratives within British Eco-Paganism, in *Popular Spiritualities: The Politics of Contemporary Enchantment*, edited by L. Hume and K. McPhillips. Aldershot: Ashgate, 175-86.

Lévy-LeBlond, J.-M. 1992. About misunderstandings about misunderstandings. *Public Understanding of Science*, 1(1), 17-21.

Lewenstein, B. 1995. Science and the media, in *Handbook of Science and Technology Studies*, edited by S. Jasanoff et al. London: Sage, 343-60.

Lewis, J. 2008. *Cultural Studies: The Basics*. 2nd Edition. London: Sage.

Lock, S.J. 2008. *Lost in Translations: Discourses, Boundaries and Legitimacy in the Public Understanding of Science in the UK*. Ph.D. Thesis. London: Department of Science and Technology Studies, UCL.

Locke, S. 1991. "Depth hermeneutics": some problems in application to the public understanding of science. *Sociology*, 25(3), 375-94.

Locke, S. 1994. The use of scientific discourse by creation scientists: some preliminary findings. *Public Understanding of Science*, 3, 403-24.

Locke, S. 1999a. *Constructing "The Beginning": Discourses of Creation Science.* Mahwah, NJ: Lawrence Erlbaum Associates.

Locke, S. 1999b. Golem science and the public understanding of science: from deficit to dilemma. *Public Understanding of Science*, 8, 75-92.

Locke, S. 2001a. *Metaphorical Power: Asymmetries in the Use of Analogy in a Case of Institutional Talk Involving a Deviant Scientist*, Orders of Ordinary Action Conference, IIEMCA, Manchester Metropolitan University, UK, 9-11 July 2001.

Locke, S. 2001b. Sociology and the public understanding of science: from rationalization to rhetoric. *British Journal of Sociology*, 52, 1-18.

Locke, S. 2001c. *Towards a Reflexive Society: Musings of a Rhetorical Sociologist on the Issue of "Economic Growth"*. Why economic growth? The meaning and measurement of GDP Conference, Kingston University, UK, 30-31 August 2001.

Locke, S. 2002. The public understanding of science – a rhetorical invention. *Science, Technology, and Human Values*, 27(1), 87-111.

Locke, S. 2004a. Charisma and the iron cage: rationalization, science and Scientology. *Social Compass*, 51(1), 111-31.

Locke, S. 2004b. Creationist discourse and the management of political-legal argumentation: comparing Britain and the USA, in *The Cultures of Creationism: Anti-Evolutionism in English-Speaking Countries*, edited by S. Coleman and L. Carlin. Aldershot: Ashgate, 45-65.

Locke, S. 2005. Fantastically reasonable: ambivalence in the representation of science and technology in super-hero comics. *Public Understanding of Science*, 14(1), 25-46.

Locke, S. 2007. *The Struggle for "Sustainability": Notes Towards a Sociological Anarchism*. Institute of Social Science Social Sustainability Mini-Conference, Kingston University, UK, 30 June 2007.

Locke, S. 2008. *The Spirit(ualities) of Science: Science and Religion in the Comic Books of Dave Sim and Alan Moore – A Preliminary Analysis*. Religion and

Youth, BSA Sociology of Religion Study Group Conference, Birmingham, UK, 8-10 April 2008.

Locke, S. 2009a. *Colouring in the "Black Box": Alternative Renderings of Scientific Visualisations in Two Comic Book Cosmologies*, Fourth Annual Science and the Public Conference, University of Brighton, UK, 13-14 June 2009.

Locke, S. 2009b. Considering comics as medium, art and culture – the case of *From Hell*. *SCAN – The Journal of Media, Art and Culture* [Online], 6(1). Available at http://scan.net.au/scan/journal/display.php?journal_id=127 [accessed: 4 March 2010].

Locke, S. 2009c. Conspiracy culture, blame culture and rationalisation. *Sociological Review*, 57(4), 567-85.

Lukes, S. 2005. *Power: A Radical View*. 2nd Edition. Basingstoke: Palgrave Macmillan.

Lynch, M. 1985. *Art and Artifact in Laboratory Science*. London: Routledge and Kegan Paul.

Lynch, M. 1992. Extending Wittgenstein: the pivotal move from epistemology to the sociology of science, in *Science as Practice and Culture*, edited by A. Pickering. Chicago: University of Chicago, 215-65.

Lynch, M. 1997. A sociology of knowledge machine. *Ethnnographic Studies*, 2, 16-38.

Lynch, M. 2006. From Ruse to farce. *Social Studies of Science*, 36(6), 819-26.

Lynch, M. 2009. Working out what Garfinkel could possibly be doing with "Durkheim's aphorism", in *Sociological Objects: Reconfigurations of Social Theory*, edited by G. Cooper, A. King and R. Rettie. Farnham: Ashgate, 101-18.

Lyne, J. and Howe, H.F. 1990. The rhetoric of expertise: E.O. Wilson and sociobiology. *Quarterly Journal of Speech*, 76, 134-51.

Lyne, J. and Howe, H.F. 1997. "Punctuated equilibria": rhetorical dynamics of a scientific controversy, in *Landmark Essays on Rhetoric of Science: Case Studies*, edited by R.A. Harris. Mahwah, NJ: Lawrence Erlbaum Associates, 69-86.

Lyotard, J.-F. 1984. *The Postmodern Condition: A Report on Knowledge*, translated by G. Bennington and B. Massumi. Manchester: Manchester University Press.

Magin, U. 2008. Horns of a dilemma. *Fortean Times*, 242, November, 55.

Main, R. 1999. Magic and science in the modern western tradition of the *I Ching*. *Journal of Contemporary Religion*, 14(2), 263-75.

Marcuse, H. 1964. *One-Dimensional Man*. London: Routledge and Kegan Paul.

Martin, B. 1981. *A Sociology of Contemporary Cultural Change*. Oxford: Blackwell.

Martin, J. and Veel, R. (eds) 1998. *Reading Science: Critical and Functional Perspectives on Discourse of Science*. London: Routledge.

McCarthy, E.D. 1996. *Knowledge as Culture: The New Sociology of Knowledge*. London: Routledge.

McCloskey, D.N. 1983. The rhetoric of economics. *Journal of Economic Literature*, 21, 481-517.
McCloud, S. 1993. *Understanding Comics: The Invisible Art*. Northampton, MA: Kitchen Sink Press.
McGee, M.C. and Lyne, J.R. 1987. What are nice folks like you doing in a place like this? Some entailments of treating knowledge claims rhetorically, in *The Rhetoric of the Human Sciences: Language and Argument in Scholarship and Public Affairs*, edited by J.S. Nelson, A. Megill and D.N. McCloskey. Madison, WI: University of Wisconsin Press, 381-406.
McGuigan, J. 1992. *Cultural Populism*. London: Routledge.
McPhillips, K. 2006. Believing in post-modernity: technologies of enchantment in contemporary Marian devotion, in *Popular Spiritualities: The Politics of Contemporary Enchantment*, edited by L. Hume and K. McPhillips. Aldershot: Ashgate, 147-58.
Mellor, F. 2003. Between fact and fiction: demarcating science from non-science in popular physics books. *Social Studies of Science*, 33(4), 509-38.
Mellor, F., Davies, S.R. and Bell, A.R. 2008. Introduction: "solverating the problematising", in *Science and its Publics*, edited by A.R. Bell, S.R. Davies and F. Mellor. Newcastle: Cambridge Scholars, 1-14.
Merton, R.K. 1968a. Science and democratic social structure, in *Social Theory and Social Structure*. 2nd Edition. New York: Free Press, 604-15.
Merton, R.K. 1968b. Science and the social order, in *Social Theory and Social Structure*. 2nd Edition. New York: Free Press, 591-603.
Merton, R.K. 1970. *Science, Technology and Society in Seventeenth Century England*. New York: Fertig.
Messeri, L.R. 2010. The problem with Pluto: conflicting cosmologies and the classification of planets. *Social Studies of Science*, 40(2), 187-214.
Michael, M. 1992. Lay discourses of science: science-in-general, science-in-particular, and self. *Science, Technology, and Human Values*, 17(3), 313-33.
Michael, M. 2000. *Reconnecting Culture, Technology and Nature: From Society to Heterogeneity*. London: Routledge.
Michael, M. 2006. *Technoscience and Everyday Life: The Complex Simplicities of the Mundane*. Maidenhead: Open University Press.
Michael, M. 2009. Publics performing publics: of PiGs, PiPs and politics. *Public Understanding of Science*, 18(5), 617-31.
Michael, M. and Birke, L. 1994. Enrolling the core set: the case of animal experimentation controversy. *Social Studies of Science*, 24(1), 81-95.
Michell, J. 1973. *The View Over Atlantis*. London: Abacus.
Michell, J. 1974. *The Flying Saucer Vision*. London: Abacus.
Milburn, C. 2008. Science from Hell: Jack the Ripper and Victorian vivisection, in *Science Images and Popular Images of the Sciences*, edited by B. Hüppauf and P. Weingart. London: Routledge, 125-57.
Miller, R. 1987. *Bare-Faced Messiah: The True Story of L. Ron Hubbard*. London: Sphere.

Miller, T. and McHoul, A. 1998. *Popular Culture and Everyday Life*. London: Sage.

Mills, C.W. 1999. *The Sociological Imagination*. New York: Oxford University Press.

Mills, S. 1997. *Discourse*. London: Routledge.

Molloy, S. 1980. Max Weber and the religions of China: any way out of the maze? *British Journal of Sociology*, 31(3), 377-400.

Moore, A. and Campbell, E. 2000. *From Hell*. London: Knockabout.

Moore, A., Williams III, J.H., Gray, M., Cox, J. and Klein, T. 2000a. Promethea: sex, stars and serpents. *Promethea* 10, November. La Jolla, CA: America's Best Comics.

Moore, A., Williams III, J.H., Gray, M., Cox, J. and Klein, T. 2000b. Promethea: pseunami. *Promethea* 11, December. La Jolla, CA: America's Best Comics.

Moore, A., Williams III, J.H., Gray, M., Cox, J. and Klein, T. 2001a. Promethea: the fields we know. *Promethea* 13, April. La Jolla, CA: America's Best Comics.

Moore, A., Williams III, J.H., Gray, M., Cox, J. and Klein, T. 2001b. Promethea: Mercury rising. *Promethea* 15, August. La Jolla, CA: America's Best Comics.

Moore, A., Williams III, J.H., Gray, M., Villarrubia, J., Cox, J. and Klein, T. 2001. Promethea: metaphore. *Promethea* 12, February. La Jolla, CA: America's Best Comics.

Moore, A., Williams III, J.H., Gray, M., Villarrubia, J., Cox, J. and Klein, T. 2004. Promethea: the radiant, heavenly city. *Promethea* 31, October. La Jolla, CA: America's Best Comics.

Moore, A., Williams III, J.H. and Klein, T. 2005. Promethea: universe. *Promethea* 32, November. La Jolla, CA: America's Best Comics.

Moores, S. 1993. *Interpreting Audiences: Ethnography of Media Consumption*. London: Sage.

Morley, D. 1980. *The "Nationwide" Audience: Structure and Decoding*. London: British Film Institute.

Mulkay, M. 1976. Norms and ideology in science. *Social Science Information*, 15(4/5), 637-56.

Mulkay, M. 1981. Action and belief or scientific discourse? A possible way of ending intellectual vassalage in social studies of science. *Philosophy of the Social Sciences*, 11, 163-71.

Mulkay, M. 1985. *The Word and the World: Explorations in the Form of Sociological Analysis*. London: Allen and Unwin.

Mulkay, M. 1993. *Science and the Sociology of Knowledge*. Aldershot: Gregg Revivals.

Mulkay, M. and Gilbert, G.N. 1982. Accounting for error: how scientists construct their social world when they account for correct and incorrect belief. *Sociology*, 16, 165-83.

Mulkay, M., Potter, J. and Yearley, S. 1983. Why an analysis of scientific discourse is needed, in *Science Observed: Perspectives on the Social Study of Science*, edited by K.D. Knorr-Cetina and M. Mulkay. London: Sage, 171-203.

Murch, R. 2009. A brief history of the ouija board. *Fortean Times*, 249, June, 32-33.

Myers, G. 1985. Nineteenth-century popularizations of thermodynamics and the rhetoric of social prophecy. *Victorian Studies*, 29, 35-66.

Myers, G. 1990. *Writing Biology: Texts in the Social Construction of Scientific Knowledge.* Madison, WI: University of Wisconsin Press.

Myers, G. 1997. Text as knowledge claims: the social construction of two biology articles, in *Landmark Essays on Rhetoric of Science: Case Studies*, edited by R.A. Harris. Mahwah, NJ: Lawrence Erlbaum Associates, 187-215.

Myers, G. and Macnaghten, P. 1998. Rhetorics of environmental sustainability: commonplaces and places. *Environment and Planning A*, 30, 333-53.

Nelkin, D. 1982. *The Creation Controversy: Science or Scripture in the Schools.* London: W.W. Norton and Co.

Nelkin, D. (ed.) 1992a. *Controversy: Politics of Technical Decisions*. 3rd Edition. London: Sage.

Nelkin, D. 1992b. Science, technology and political conflict: analyzing the issues, in *Controversy: Politics of Technical Decisions*, edited by D. Nelkin. 3rd Edition. London: Sage, ix-xxv.

Nelkin, D. and Lindee, M.S. 1995. *The DNA Mystique: The Gene as a Cultural Icon.* New York: W.H. Freeman and Co.

Nelson, J.S. 1987. Seven rhetorics of inquiry: a provocation, in *The Rhetoric of the Human Sciences: Language and Argument in Scholarship and Public Affairs*, edited by J.S. Nelson, A. Megill and D.N. McCloskey. Madison, WI: University of Wisconsin Press, 407-34.

Nelson, J.S., Megill, A. and McCloskey, D.N. (eds) 1987. *The Rhetoric of the Human Sciences: Language and Argument in Scholarship and Public Affairs.* Madison, WI: University of Wisconsin Press.

Nixon, N. 1999. They're not all lunatics on the fringe. *Fortean Studies,* 6, 163-77.

Nowotny, H., Scott, P. and Gibbons, M. 2001. *Re-thinking Science: Knowledge and the Public in an Age of Uncertainty*. Cambridge: Polity.

Nyberg, A.K. 1998. *Seal of Approval: The History of the Comics Code*. Jackson: University of Mississippi Press.

Nyhart, L.K. 1991. Writing zoologically: The *Zeitschrift für wissenschaftliche Zoologie* and the zoological community in late nineteenth-century Germany, in *The Literary Structure of Scientific Argument: Historical Studies*, edited by P. Dear. Philadelphia: University of Pennsylvania Press, 43-71.

Odell, R. 2006. *Ripperology: A Study of the World's First Serial Killer and a Literary Phenomenon.* Kent, OH: Kent State University Press.

Oehlert, M. 1995. From Captain America to Wolverine: cyborgs in comic books, alternative images of cybernetic heroes and villains, in *The Cyborg Handbook*, edited by C.H. Gray. London: Routledge, 219-32.

Overton, W.R. 1982. Memorandum opinion, in *The Creation Controversy: Science or Scripture in the Schools*, by D. Nelkin. London: W.W. Norton and Co, 201-28.

Pahl, R.E. 1984. *Divisions of Labour*. Oxford: Blackwell.

Parish, J. 2001. The age of anxiety, in *The Age of Anxiety: Conspiracy Theory and the Human Sciences*, edited by J. Parish and M. Parker. Oxford: Blackwell, 1-16.

Parish, J. and Parker, M. (eds) 2001. *The Age of Anxiety: Conspiracy Theory and the Human Sciences*. Oxford: Blackwell.

Park, H.J. 2001. The creation-evolution debate: carving creationism in the public mind. *Public Understanding of Science*, 10(2), 173-86.

Park, R.L. 2000. *Voodoo Science: The Road from Foolishness to Fraud.* New York: Oxford University Press.

Parker, M. 1995. Working together, working apart: management culture in a manufacturing firm. *Sociological Review*, 43, 518-547.

Parker, M. 2001. Human science as conspiracy theory, in *The Age of Anxiety: Conspiracy Theory and the Human Sciences*, edited by J. Parish and M. Parker. Oxford: Blackwell, 191-207.

Parkin, F. 1982. *Max Weber*. London: Routledge.

Parrinder, P. 1980. *Science Fiction: Its Criticism and Teaching.* London: Methuen.

Parsons, T. 1968. *The Structure of Social Action: A Study in Social Theory with Special Reference to a Group of European Writers. Volume I.* New York: The Free Press.

Partridge, C. 2004. *The Re-Enchantment of the West, Volume 1: Alternative Spiritualities, Sacralization, Popular Culture and Occulture.* London: T&T Clark International.

Pauwels, L. and Bergier, J. 2001. *The Morning of the Magicians*, translated by R. Myers. London: Souvenir Press.

Perelman, C.H. and Olbrechts-Tyteca, L. 1969. *The New Rhetoric: A Treatise on Argumentation*, translated by J. Wilkinson and P. Weaver. Notre Dame: University of Notre Dame Press.

Perez, D. 2005. In defence of the Patterson-Gimlin film. *Fortean Times*, 192, January, 36-37.

Picart, C.J.S. 1994. Scientific controversy as farce: the Benveniste-Maddox counter trials. *Social Studies of Science*, 24(1), 7-37.

Pigden, C. 1995. Popper revisited, or what is wrong with conspiracy theories? *Philosophy of the Social Sciences*, 25(1), 3-34.

Pilkington, M. 2009. Fortean Bureau of Investigation. *Fortean Times*, 246, March, 40-42.

Pipes, D. 1997. *Conspiracy: How the Paranoid Style Flourishes and Where it Comes from*. Boston: Free Press.

Poggi, G. 2006. *Weber: A Short Introduction.* Cambridge: Polity.

Pollner, M. 1974. Mundane reasoning. *Philosophy of the Social Sciences*, 4, 35-54.

Pollner, M. 1975. "The very coinage of your brain": the anatomy of reality disjunctures. *Philosophy of the Social Sciences*, 5, 411-30.

Pollner, M. 1987. *Mundane Reason: Reality in Everyday and Sociological Discourse.* Cambridge: Cambridge University Press.

Pollner, M. 1991. Left of ethnomethodology: the rise and decline of radical reflexivity. *American Sociological Review*, 56, 370-80.

Pomerantz, A. 1986. Extreme case formulations: a way of legitimizing claims. *Human Studies*, 9, 219-29.

Popper, K. 1966. *The Open Society and its Enemies (2 volumes)*. 5th Edition. London: Routledge and Kegan Paul.

Popper, K. 1972. *Objective Knowledge.* Oxford: Clarendon Press.

Popper, K. 2002. *The Logic of Scientific Discovery.* London: Routledge.

Possamai, A. 2006. Superheroes and the development of latent abilities: a hyper-real re-enchantment?, in *Popular Spiritualities: The Politics of Contemporary Enchantment*, edited by L. Hume and K. McPhillips. Aldershot: Ashgate, 53-62.

Potter, J. 1996. *Representing Reality: Discourse, Rhetoric and Social Construction.* London: Sage.

Potter, J. and Wetherell, M. 1987. *Discourse and Social Psychology: Beyond Attitudes and Behaviour.* London: Sage.

Potter, J., Wetherell, M. and Chitty, A. 1991. Quantification rhetoric – cancer on television. *Discourse and Society*, 2, 333-65.

Prelli, L.J. 1989a. *A Rhetoric of Science: Inventing Scientific Discourse.* Columbia, SC: University of South Carolina Press.

Prelli, L.J. 1989b. The rhetorical construction of scientific ethos, in *Rhetoric in the Human Sciences*, edited by H.W. Simons. London: Sage, 48-68.

Propp, V. 1968. *Morphology of the Folktale*, translated by L. Scott. 2nd Edition. Austin: University of Texas Press.

Propp, V. 1984. *Theory and History of Folklore*, translated by A.Y. Martin and R.P. Martin. Minneapolis, MN: University of Minnesota.

Pustz, M.J. 1999. *Comic Book Culture: Fanboys and True Believers*. Jackson: University Press of Mississippi.

Randles, J. 2005. Rendlesham: evolving theories. *Fortean Times*, 204, December, 38-39.

Randles, J. and Clarke, D. 2005. Down to earth… *Fortean Times*, 204, December, 32-39.

Raphael, J. and Spurgeon, T. 2003. *Stan Lee and the Rise and Fall of the American Comic Book.* Chicago: Chicago Review Press.

Rawls, A. 2002. Editor's introduction, in *Ethnomethodology's Program: Working out Durkheim's Aphorism*, by H. Garfinkel. Lanham: Rowman and Littlefield, 1-64.

Rawls, A. 2009. Communities of practice vs traditional communities: the state of sociology in a context of globalization, in *Sociological Objects: Reconfigurations of Social Theory*, edited by G. Cooper, A. King and R. Rettie. Farnham: Ashgate, 81-99.

Reeves, B. 1999. Who invented the wheel? The motor of life. *CSM Pamphlet 323*, March. Portsmouth: Creation Science Movement.

Reeves, C. 1997. Owning a virus: the rhetoric of scientific discovery accounts, in *Landmark Essays on Rhetoric of Science: Case Studies*, edited by R.A. Harris. Mahwah, NJ: Lawrence Erlbaum Associates, 151-65.

Reinel, B. 1999. Reflections on cultural studies of technoscience. *European Journal of Cultural Studies*, 2(2), 163-89.

Restivo, S.P. 1978. Parallels and paradoxes in modern physics and eastern mysticism I. *Social Studies of Science*, 8, 143-81.

Restivo, S.P. 1982. Parallels and paradoxes in modern physics and eastern mysticism II. *Social Studies of Science*, 12, 37-71.

Reynolds, R. 1992. *Superheroes: A Modern Mythology*. London: Batsford.

Rhodes, M. 2000. Wonder Woman and her disciplinary powers: the queer intersection of scientific authority and mass culture, in *Doing Science and Culture*, edited by R. Reid and S. Traweek. London: Routledge, 95-118.

Richards, E. and Ashmore, M. (eds) 1996. *Social Studies of Science Special Issue*, 26(2).

Rickard, B. 2009. Outbreak! *Fortean Times*, 253, September, 30-37.

Rickard, B. and Michell, J. 2000. *Unexplained Phenomena: A Rough Guide Special*. London: Rough Guides.

Rickard, B., Lachman, G., Nicholson, J., Harpur, P., Brown, A. and Marshall, S. 2009. John Michell: a modern Merlin. *Fortean Times*, 249, June, 38-49.

Roberts, A. 2004. The space baby. *Fortean Times*, 191, Special, 32-38.

Roberts, A. and Clarke, D. 2009a. Flying saucery. *Fortean Times*, 246, March, 28.

Roberts, A. and Clarke, D. 2009b. Flying saucery. *Fortean Times*, 255, November, 28.

Rosevear, D. 1991. *Creation Science: Confirming that the Bible is Right*. Chichester: New Wine Press.

Ross, A. (ed.) 1996. *Science Wars*. Durham, NC: Duke University Press.

Royal Society, 1985. *The Public Understanding of Science*. London: Royal Society.

Rubinstein, W.D. 2000. The hunt for Jack the Ripper. *History Today*, 50(5), 10-19.

Rumbelow, D. 2004. *The Complete Jack the Ripper: Fully Revised and Updated*. Harmondsworth: Penguin.

Sacks, H. 1972. An initial investigation of the usability of conversational data for doing sociology, in *Studies in Social Interaction*, edited by D. Sudnow. New York: The Free Press, 31-74.

Sacks, H. 1974. On the analysability of stories by children, in *Ethnomethodology: Selected Readings*, edited by R. Turner. Harmondsworth: Penguin, 216-32.

Sacks, H. 1984. On doing "being ordinary", in *Structures of Social Action: Studies in Conversation Analysis*, edited by J.M. Atkinson and J. Heritage. Cambridge: Cambridge University Press, 413-29.

Sacks, H. 1995a. *Lectures on Conversation Volume 1*. Oxford: Blackwell.

Sacks, H. 1995b. *Lectures on Conversation Volume 2*. Oxford: Blackwell.

Sacks, H., Schegloff, E.A. and Jefferson, G. 1974. A simplest systematics for the organisation of turn-taking in conversation. *Language*, 50, 696-735.

Saler, M. 2003. "Clap if you believe in Sherlock Holmes": mass culture and the re-enchantment of modernity, c.1890-c. 1940. *The Historical Journal*, 46(3), 599-622.

Schegloff, E. 1995. Introduction, in *Lectures on Conversation Volume 1*, by H. Sacks. Oxford: Blackwell, ix-lxii.

Schegloff, E.A. 1997. Whose text? Whose context? *Discourse and Society*, 8, 165-87.

Schegloff, E.A. 1998. Reply to Wetherell. *Discourse and Society*, 9(3), 457-60.

Schegloff, E.A. 1999a. "Schegloff's texts" as "Billig's data": a critical reply. *Discourse and Society*, 10(4), 558-72.

Schegloff, E.A. 1999b. Naivete vs sophistication or discipline vs self indulgence. *Discourse and Society*, 10(4), 577-82.

Schegloff, E.A. 2007. A tutorial on membership categorization. *Journal of Pragmatics*, 39, 462-82.

Schroeder, R. 1995. Disenchantment and its discontents: Weberian perspectives on science and technology. *Sociological Review*, 43(2), 227-50.

Schutz, A. 1972. *The Phenomenology of the Social World*, translated by G. Walsh and F. Lehnert. London: Heinemann.

Scott, J. 1991. *Who Rules Britain?* Cambridge: Polity.

Shafir, G. 1985. The incongruity between destiny and merit: Max Weber on meaningful existence and modernity. *British Journal of Sociology*, 36(4), 516-30.

Shapin, S. 1990. Science and the public, in *Companion to the History of Modern Science*, edited by R.C. Olby et al. London: Routledge, 990-1007.

Shapin, S. 1994. *A Social History of Truth: Civility and Science in Seventeenth-Century England*. Chicago: University of Chicago Press.

Shapin, S. and Schaffer, S. 1985. *Leviathan and the Air-Pump: Hobbes, Boyle, and the Experimental Life.* Princeton, NJ: Princeton University Press.

Sharrock, W.W. 1974. On owning knowledge, in *Ethnomethodology: Selected Readings*, edited by R.Turner. Harmondsworth: Penguin, 45-53.

Sheldrake, R. 1994. *Seven Experiments That Could Change the World: A Do-It-Yourself Guide to Revolutionary Science*. London: Fourth Estate.

Shelley, M. 1994. *Frankenstein, Or the Modern Prometheus: The 1818 Text.* Oxford: Oxford University Press.

Shermer, M. 1998. *Why People Believe Weird Things: Pseudoscience, Superstition and Other Confusions of Our Time*. New York: W.H. Freeman and Co.

Siegel, J. and Shuster, J. 1997[1938]. Superman, champion of the oppressed, in *Action Comics Archives Volume 1.* New York: DC Comics, 11-23.

Sieveking, P. 2006. Reply to Jerome Clark. *Fortean Times*, 210, June, 73.

Silverman, D. 1998. *Harvey Sacks: Social Science and Conversation Analysis*. Cambridge: Polity.

Simmons, I. 2006. On the origin of the specious. *Fortean Times*, 211, Special, 46-50.

Simons, H.W. (ed.) 1989. *Rhetoric in the Human Sciences*. London: Sage.

Skal, D. 1998. *Screams of Reason: Mad Science and Modern Culture*. New York: W.W. Norton.

Skinner, Q. (ed.) 1990. *The Return of Grand Theory in the Human Sciences*. Cambridge: Cambridge University Press.

Smith, D. 1978. "K is mentally ill": the anatomy of a factual account. *Sociology*, 12, 23-53.

Smith, J.M. 1997. Jack the Ripper, Sherlock Holmes, and the New Woman. *News From Nowhere*, 2, 121-36.

Smith, M. 1998. *Social Science in Question*. London: Sage.

Snow, C.P. 1964. *The Two Cultures and A Second Look*. 2nd Edition. Cambridge: Cambridge University Press.

Spark, A. 2001. Conjuring order: the New World Order and conspiracy theories of globalization, in *The Age of Anxiety: Conspiracy Theory and the Human Sciences*, edited by J. Parish and M. Parker. Oxford: Blackwell, 46-62.

Steinmeyer, J. 2008. *Charles Fort: The Man Who Invented the Supernatural*. London: William Heinemann.

Steranko, J. 1970. *The Steranko History of Comics Volume One*. Reading, PA: Supergraphics.

Stevenson, I. 2009. Table-turning. *Fortean Times*, 254, October, 77.

Stevenson, N. 2002. *Understanding Media Cultures: Social Theory and Mass Communications*. 2nd Edition. London: Sage.

Storey, J. 2003. *Inventing Popular Culture: From Folklore to Globalization*. Oxford: Blackwell.

Sugden, P. 2002. *The Complete History of Jack the Ripper*. New Edition. London: Robinson.

Sutcliffe, S.J. 2006. Practising New Age soteriologies in the rational order, in *Popular Spiritualities: The Politics of Contemporary Enchantment*, edited by L. Hume and K. McPhillips. Aldershot: Ashgate, 159-74.

Swedberg, R., Himmelstrand, U. and Brulin, G. 1987. The paradigm of economic sociology. *Theory and Society*, 16, 169-213.

Tatar, M. 1995. *Lustmord: Sexual Murder in Weimar Germany*. Princeton, NJ: Princeton University Press.

Taylor, C.A. 1992. Of audience, expertise and authority: the evolving creationism debate. *Quarterly Journal of Speech*, 78, 277-95.

Taylor, C.A. 1996. *Defining Science: A Rhetoric of Demarcation*. Madison, WI: University of Wisconsin Press.

Thomas, R. and Buscema, J. 1972. Godhood's end! *The Avengers*, 97, March. New York: Marvel Comics Group.

Thomas, R. and Buscema, S. 1971. The only *good* alien.... *The Avengers*, 89, June. New York: Marvel Comics Group.

Tiryakian, E.A. 1992. Dialectics of modernity: reenchantment and dedifferentiation as counterprocesses, in *Social Change and Modernity*, edited by H. Haferkamp and N.J. Smelser. Berkeley, CA: University of California Press, 78-94.

Toumey, C.P. 1992. The moral character of mad scientists: a cultural critique of science. *Science, Technology, and Human Values*, 17, 411-37.

Toumey, C.P. 1994. *God's Own Scientists: Creationists in a Secular World.* New Brunswick, NJ: Rutgers University Press.

Truzzi, M. 2006. On pseudo-skepticism: a commentary. *Fortean Times*, 208, April, 59.

Tudor, A. 1989. *Monsters and Mad Scientists: A Cultural History of the Horror Movie*. Oxford: Blackwell.

Tumminia, D. 1998. How prophecy never fails: interpretive reason in a flying-saucer group. *Sociology of Religion*, 59(2), 157-70.

Turner, B.S. 1987. State, science and economy in traditional societies: some problems in Weberian sociology of science. *British Journal of Sociology*, 38(1), 1-23.

Turney, J. 1998a. *Frankenstein's Footsteps: Science, Genetics and Popular Culture.* New Haven: Yale University Press.

Turney, J. 1998b. *To Know Science is to Love it? Observations From Public Understanding of Science Research*. London: COPUS.

van Dijck, J. 1998. *Imagenation: Popular Images of Genetics.* Basingstoke: Macmillan.

Varnum, R. and Gibbons, C.T. 2001. *The Language of Comics: Word and Image.* Jackson: University Press of Mississippi.

Veitch, R. 1985-1986. *The One*, 1 – 6, July 1985 – May 1986. New York: Marvel Comics Group.

Vieth, E. 2001. *Screening Science: Contexts, Texts, and Science in Fifties Science Fiction Films.* Lanham, MD: Scarecrow Press.

Vološinov, V.N. 1973. *Marxism and the Philosophy of Language*, translated by L. Matejka and I.R. Titunik. Cambridge, MA: Harvard University Press.

von Daniken, E. 1971. *Chariots of the Gods?*, translated by M. Heron. London: Corgi.

Waddell, C. 1994. Rhetoric of environmental policy: from critical practice to the social construction of theory. *Social Epistemology*, 8, 289-310.

Waddell, C. 1997. The role of *pathos* in the decision-making process: a study in the rhetoric of science policy, in *Landmark Essays on Rhetoric of Science: Case Studies*, edited by R.A. Harris. Mahwah, NJ: Lawrence Erlbaum Associates, 127-49.

Waid, M., Wieringo, M. and Kesel, K. 2003. Unthinkable Parts 1-4. *Fantastic Four*, 68/497-71/500, June-September. New York: Marvel Comics.

Walkowitz, J. 1992. *City of Dreadful Delight: Narratives of Sexual Danger in Late-Victorian London*. London: Virago.

Wallis, R. 1976. *The Road to Total Freedom: A Sociological Analysis of Scientology.* London: Heinemann.

Wallis, R. (ed.) 1979. *On the Margins of Science: The Social Construction of Rejected Knowledge*. Keele: Keele University Press.

Warwick, A. and Willis, M. (eds) 2007. *Jack the Ripper: Media, Culture, History*. Manchester: Manchester University Press.

Watkins, A. 1974[1925]. *The Old Straight Track*. London: Abacus.

Watson, R. 1997. Some general reflections on "categorization" and "sequence" in the analysis of conversation, in *Culture in Action: Studies in Membership Categorization Analysis*, edited by S. Hester and P. Eglin. Washington DC: International Institute for Ethnomethodology and Conversation Analysis and University Press of Washington, 49-75.

Weber, M. 1948a. Religious rejections of the world and their directions, in *From Max Weber: Essays in Sociology*, edited by H.H. Gerth and C.W. Mills. London: Routledge and Kegan Paul, 323-59.

Weber, M. 1948b. Science as a vocation, in *From Max Weber: Essays in Sociology*, edited by H.H. Gerth and C.W. Mills. London: Routledge and Kegan Paul, 129-56.

Weber, M. 1948c. The social psychology of the world religions, in *From Max Weber: Essays in Sociology*, edited by H.H. Gerth and C.W. Mills. London: Routledge and Kegan Paul, 267-301.

Weber, M. 1949. *The Methodology of the Social Sciences*, translated by E.A. Shils and H.A. Finch. New York: Free Press.

Weber, M. 1951. *The Religion of China: Confucianism and Taoism*, translated by H.H. Gerth. New York: Free Press.

Weber, M. 1952. *Ancient Judaism*, translated by H.H. Gerth and D. Martindale. New York: Free Press.

Weber, M. 1958. *The Religion of India*, translated by H.H. Gerth and D. Martindale. New York: Free Press.

Weber, M. 1965. *The Sociology of Religion*, translated by T. Parsons. London: Methuen.

Weber, M. 1968. *Economy and Society: An Outline of Interpretive Sociology*, translated by E. Fischoff et al. New York: Bedminster Press.

Weber, M. 1976. *The Protestant Ethic and the Spirit of Capitalism*, translated by T. Parsons. London: Unwin.

Weingart, P., Muhl, C. and Pansegrau, P. 2003. Of power maniacs and unethical geniuses: science and scientists in fiction film. *Public Understanding of Science*, 12, 279-87.

Weizenbaum, J. 1984. *Computer Power and Human Reason: From Judgement to Calculation*. Harmondsworth: Penguin.

Wertham, F. 1955. *Seduction of the Innocent*. London: Museum Press.

Westrum, R. 1977. Social intelligence about anomalies: the case of UFOs. *Social Studies of Science*, 7, 271-302.

Wetherell, M. 1998. Positioning and interpretative repertoires: conversation analysis and post-structuralism in dialogue. *Discourse and Society*, 9(3), 387-412.

Whalen, M. and Zimmerman, D. 1990. Describing trouble: practical epistemology in citizen calls to the police. *Language in Society*, 19, 465-92.

Wheen, F. 2004. *How Mumbo-Jumbo Conquered the World: A Short History of Modern Delusions*. London: Fourth Estate.

White, J.B. 1987. The arts of cultural and communal life, in *The Rhetoric of the Human Sciences: Language and Argument in Scholarship and Public Affairs*, edited by J.S. Nelson, A. Megill and D. McCloskey. Madison, WI: University of Wisconsin Press, 298-318.

Whitehead, H. 1974. Reasonably fantastic: some perspectives on Scientology, science fiction and occultism, in *Religious Movements in Contemporary America*, edited by I.I. Zaretsky and M.P. Leone. Princeton, NJ: Princeton University Press, 547-87.

Whitehead, H. 1987. *Renunciation and Reformulation: A Study of Conversion in an American Sect*. Ithaca: Cornell University Press.

Widdicombe, S. 1998. "But you don't class yourself": the interactional management of category membership and non-membership, in *Identities in Talk*, edited by C. Antaki and S. Widdicombe. London: Sage, 52-70.

Widdicombe, S. and Wooffitt, R. 1995. *The Language of Youth Subcultures: Social Identity in Action*. Hemel Hempstead: Harvester Wheatsheaf.

Williams, R. 1976. *Keywords: A Vocabulary of Culture and Society*. London: Fontana.

Willis, P. 1977. *Learning to Labour: How Working Class Kids Get Working Class Jobs*. Farnborough: Saxon House.

Wilson, B. 1966. *Religion in a Secular Society: A Sociological Comment*. London: C.A. Watts.

Wilson, B. 1970. *Religious Sects: A Sociological Study*. New York: McGraw-Hill.

Wilson, B. 1976. *Contemporary Transformations of Religion*. London: Oxford University Press.

Wilson, B. 1990. Scientology: a secularized religion, in *The Social Dimensions of Sectarianism: Sects and New Religious Movements in Contemporary Society*. Oxford: Clarendon Press, 267-88.

Wilson, C. 1979. *Mysteries: An Investigation into the Occult, the Paranormal and the Supernatural*. London: Granada.

Wilson, C. and Odell, R. 1987. *Jack the Ripper: Summing Up and Verdict*. London: Corgi.

Wilson, E.O. 2000. *Sociobiology: The New Synthesis*. 25th Anniversary Edition. Cambridge, MA: Harvard University Press.

Wilson, S. 2008. Laughing with Charles Fort. *Fortean Times*, 242, November, 56-57.

Winch, P. 1990. *The Idea of a Social Science and its Relation to Philosophy*. London: Routledge.

Winiecki, D. 2008. The expert witness and courtroom discourse: applying micro and macro forms of discourse analysis to study process and the "doings of doings" for individuals and society. *Discourse and Society*, 19(6), 765-81.

Wolpert, L. 1992. *The Unnatural Nature of Science.* London: Faber and Faber.

Wolverton, M. 2002. *The Science of Superman.* New York: ibooks.

Woodlief, A.M. 1981. Science in popular culture, in *Handbook of American Popular Culture Volume 3*, edited by M.T. Inge. Westport, CT: Greenwood, 429-58.

Wooffitt, R. 1992. *Telling Tales of the Unexpected: The Organization of Factual Discourse.* Hemel Hempstead: Harvester Wheatsheaf.

Wooffitt, R. 2005. *Conversation Analysis and Discourse Analysis: A Comparative and Critical Introduction.* London: Sage.

Wooffitt, R. 2006. *The Language of Mediums and Psychics: The Social Organization of Everyday Miracles.* Aldershot: Ashgate.

Woolgar, S. 1988. *Science: The Very Idea.* Chichester: Ellis Horwood and London: Tavistock.

Woolgar, S. 2002a. Five rules of virtuality, in *Virtual Society? Technology, Cyberbole, Reality*, edited by S. Woolgar. Oxford: Oxford University Press, 1-22.

Woolgar, S. (ed.) 2002b. *Virtual Society? Technology, Cyberbole, Reality*. Oxford: Oxford University Press.

Wrong, D. 1961. The oversocialized conception of man in modern sociology. *American Sociological Review*, 26, 183-93.

Wynne, B. 1991. Knowledges in context. *Science, Technology, and Human Values*, 16, 111-21.

Wynne, B. 1992. Public understanding of science research: new horizons or hall of mirrors? *Public Understanding of Science*, 1, 37-43.

Wynne, B. 1995. Public understanding of science, in *Handbook of Science and Technology Studies*, edited by S. Jasanoff et al. London: Sage, 361-88.

Wynne, B. 1996. May the sheep safely graze? A reflexive view of the expert-lay knowledge divide, in *Risk, Environment and Modernity: Towards a New Ecology*, edited by S. Lash, B. Szerszynski and B. Wynne. London: Sage, 44-83.

Wynne, B. 2003. Seasick on the Third Wave? Subverting the hegemony of propositionalism: response to Collins and Evans. *Social Studies of Science*, 33(3), 401-17.

Wynne, B. 2008. Elephants in the rooms where publics encounter "science"?: A response to Darrin Durant, "Accounting for expertise: Wynne and the autonomy of the lay public". *Public Understanding of Science*, 17(1), 21-33.

Yaco, L. and Haber, K. 2000. *The Science of the X-Men.* New York: BP Books.

Yearley, S. 1985. Vocabularies of freedom and resentment: a Strawsonian perspective on the nature of argumentation in science and the law. *Social Studies of Science*, 15, 99-126.

Yearley, S. 1994. Understanding science from the perspective of the sociology of scientific knowledge: an overview. *Public Understanding of Science*, 3, 245-58.

Yearley, S. 2005. *Making Sense of Science: Understanding the Social Study of Science*. London: Sage.

York, M. 1995. *The Emerging Network: A Sociology of the New Age and Neo-Pagan Movements*. Lanham, MD: Rowman and Littlefield.

Zappen, J.P. 1994. The rhetoric of science and the challenge of post-liberal democracy. *Social Epistemology*, 8, 261-71.

Zdenek, S. 1998. *Encoding the Illusion: Rhetoric and the Public Understanding of Science*. The Rhetoric of Science and the Rhetorical Tradition Preconvention Conference, NCA Convention, New York, 21 November 1998.

Ziman, J. 1968. *Public Knowledge: An Essay Concerning the Social Dimension of Science*. Cambridge: Cambridge University Press.

Ziman, J. 1991. Public understanding of science. *Science, Technology, and Human Values*, 16, 99-105.

Ziman, J. 2000. *Real Science: What it Is and What it Means*. Cambridge: Cambridge University Press.

Zimmerman, D.H. 1971. The practicalities of rule use, in *Understanding Everyday Life: Toward the Reconstruction of Sociological Knowledge*, edited by J.D. Douglas. London: Routledge and Kegan Paul, 221-38.

Zimmerman, D.H. and Pollner, M. 1971. The everyday world as a phenomenon, in *Understanding Everyday Life: Toward the Reconstruction of Sociological Knowledge*, edited by J.D. Douglas. London: Routledge and Kegan Paul, 80-103.

Index